On the Wings of the Storm

By

Captain Richard Duncan, M.C.

Edited By

John S. Haller, Jr.

On the Wings of the Storm

By

Captain Richard Duncan, M.C.

Edited By

John S. Haller, Jr.

ISBN: 978-1-953201-88-1

Design and layout: Jack Herris
Cover design: Aaron Weaver
Cover photo: Colorized by Aaron Weaver
Digital photo editing: Jack Herris

Aeronaut Books

Books for Enthusiasts by Enthusiasts
www.aeronautbooks.com

Table of Contents

Acknowledgements

Ralph E. McCoy, former dean of libraries and professor of journalism from 1955 until his retirement in 1976 at Southern Illinois University at Carbondale, built one of the premier collections in the United States. From books and manuscripts on the Irish renaissance and theater, to modern philosophy and freedom of the press, to diaries and correspondence from soldiers in the Civil War, World War I, and World War II, he managed to bring together under one roof some of the finest works in the English-speaking world. Today, the management of this collection is in the able hands of a resourceful staff that includes Aaron Lisec, Nicholas Guardiano, Tony Bittle, Anne Marie Hamilton-Brehm, Matt Gorzalski, and Walter Ray. My appreciation extends to each of them for their help and attentiveness to my research needs, and especially to Aaron Lisec who made me aware of this unpublished manuscript. I am also appreciative to JSTOR, ProQuest, WorldCat, Google Books, and the Hathi Trust for their facilitation of books, articles, and other essential materials needed to provide context to the manuscript. My thanks go as well to the weekly banter from my colleagues David Werlich, Howard Allen, John Dotson, David Wilson, and James Allen; and to my wife, Robin, whose critical eye and caring nature have made each of my books a useful tool and roadmap for the discerning reader.

Editors Comments

While working through a trove of diaries and correspondence from World War I soldiers in a collection located at Southern Illinois University Carbondale, I was apprised by one of the library's archivists of an unpublished memoir by Captain Richard Duncan, M.C. The memoir, titled "On the Wings of the Storm," revealed that when Great Britain declared war on Germany on August 4, 1914, Duncan was a member of the 99th Manitoba Rangers and working with a surveying party in the Thompson River Valley in British Columbia. On February 10, 1915, he was examined by a medical doctor in Prince Albert, Saskatchewan, and considered fit for duty in the Canadian Over-Seas Expeditionary Force. In his Attestation Paper for Enlistment, he stated that he was born September 3, 1888, in Winnipeg, Manitoba, that he was 5 ft. 8.5 inches in height, had fair complexion, grey eyes, light brown hair, and claimed affiliation with the Church of England. He also identified his brother, W. G. Duncan, who worked for the Thornton Woolen Mills in Moscow, Russia, as his next-of-kin.

Written late in his life, the memoir begins with Duncan serving as a lieutenant with the 1st Canadian Mounted Rifles in the Ypres Salient before receiving a transfer to the School of Aeronautics in England. There he became a Flying Officer with No. 5 Corps Squadron, serving from December 1916 to November 1917. In April 1918, he was transferred to No.106 Squadron and promoted to Captain. The memoir consists mainly of his recollections as a Corps Squadron Pilot (as distinct from the fighter or combat pilot). These pilots were the eyes and ears of the Army Corps as artillery observers; strafing enemy trenches; providing vertical, oblique, and stereo photography of the German front line; dropping spies behind the German lines; and maintaining contact during battle between the H.Q. and the troops on the ground. The responsibilities and the hazards he experienced as a Corps Squadron Pilot were greater by far than most. Not many of these men returned. Duncan was one of the few who did, and he received the Military Cross for his gallantry.

The memoir also conveys Duncan's disdain for the British class system as evinced by those officers from elite schools whose affected drawl and air of superiority conflicted with his own more egalitarian views. Referring to one particular airman in his squadron: "[He] was almost human but I had a hell of a time understanding what he was talking about. Oxford men clip their words and leave sentences hanging in the air that only they can grasp."

After the war, Duncan worked in Canada and the United States as a licensed pilot and instructor, and authored numerous books on flying including *Practical Air Navigation and Meteorology* (1929), *Simplified Aviation* (1929), and *Aircraft Flight Instructor* (1938). Presumably he also performed as a stunt pilot during those years as well since he authored the popular book *Stunt Flying* in 1930. I mention these books because he writes well and is expressive and detailed in his reminiscences of the war. In 1929, Duncan was working as a flight instructor for Curtis Flying Service at Kansas City Airport in Missouri. That same year, he married Truda Marr, a New York actress. Throughout the 1930s, Duncan continued to provide flight instruction and was a frequent speaker at aviation clubs around the country. He officially immigrated to the United States in 1959 and died in October 1966.

Southern Illinois University Carbondale purchased the manuscript from Between the Covers, a rare books dealer in New Jersey, who obtained it from Jay Garon-Brook Associates, Inc., a New York literary agency at 415 Central Park West in New York City, the same agency that handled author John Grisham's first seven novels. The manuscript came without restrictions, and except for minor editorial changes, the use of explanatory footnotes where appropriate, and the inclusion of photos taken by Duncan and his colleagues, as well as some drawn from Wikimedia Commons to provide additional context, the memoir is original and complete.

Preface

The Corps Squadron Pilot

To understand the difference between a corps squadron pilot and the fighter (combat) pilot, once explained is comparatively simple. The fighter pilot was trained for one purpose only, that of destroying enemy aircraft and in particular their combat planes. To qualify for this type of work required he be utterly fearless, skilled in acrobatics, an expert machine gunner, and ready and willing to fight against all odds. His was the most spectacular of all jobs in the air as he scoured the skies for enemy planes and engaged them in mortal combat. In doing so he was offering, though indirectly, protection to the slow-moving corps squadron planes thousands of feet below. He flew high at altitudes varying from 10,000/20,000 feet. In his hands rested the supremacy of the air. When it was held, the losses suffered by the corps squadrons were materially reduced. When the enemy were supreme the results were devastating and the odds appalling.

The corps squadron pilot was equipped to fight and repeatedly did so, but only when attacked. He was the deadliest and most versatile of all pilots, carrying a tremendous responsibility and the lives of thousands of men in his hands. As the eyes and ears of the Army Corps with which he cooperated, he prepared the way prior to battle and maintained contact with the troops on the ground until it was completed. What his duties were and how they were carried out is described in the text. His responsibilities and the hazards he confronted were greater by far than that of any other individual in the air or on the ground. Many went out but few returned—I was one of the few.

Above: Capt. Richard Duncan, M.C., during the war.

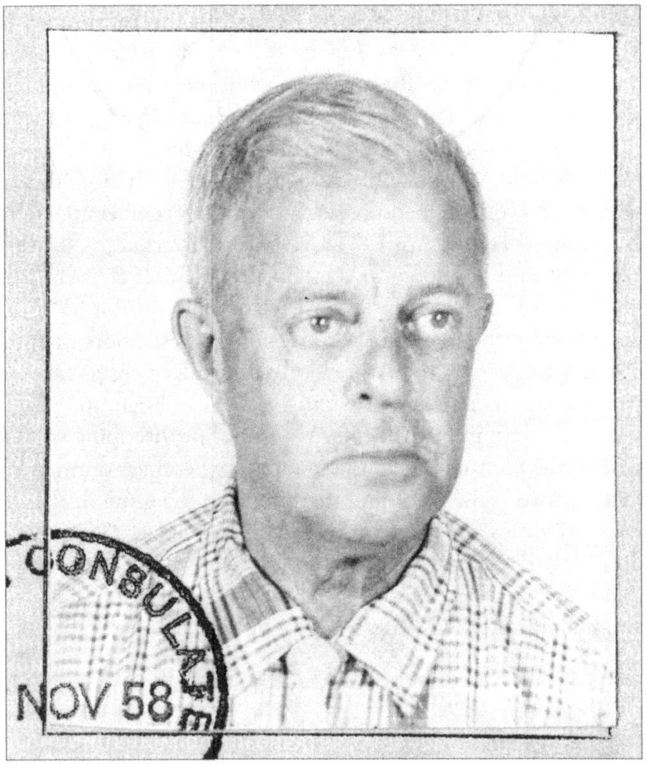

Above: Richard Duncan in later years.

Chapter 1: A Refugee from the Trenches

Above: Blériot 11 trainer. Courtesy of Aeronaut.

Above: Caudron G.3 trainer. Courtesy of Aeronaut.

On a hot summer's day in July 1916, I was sitting on the fire-step of a trench in the Ypres Salient thinking about nothing. Overhead a lark was singing a song of joy and happiness as he hovered in the clear blue sky, high above this desolate and forbidding land. It was deathly still and I was alone.

My regiment, the 1st Canadian Mounted Rifles (C.M.R.) from Western Canada occupied the Apex, the pointed triangular strip of Belgium soil, honey-combed with shallow and narrow trenches hastily dug after a tough and bloody battle a few weeks before and were little more than ditches, in which one, in the daytime, had to walk crouching in a stoop. Facing us across the other side was "No Man's Land," a once beautiful woodland area known as Sanctuary Woods, a favorite summer retreat through the centuries for the long-gone citizens of that ancient city of Ypres – or "Wipers" as the Tommies called it. There the citizens held their picnics and other affairs. Ravaged by two years of war, all that was now left were the scarred and shell torn trunks of thousands of magnificent oak, beech, and walnut trees reaching into the sky like countless pilings, stripped clean of branch and twig, and the floor below swept bare of underbrush by shell and fire.

By night it was a weird and ghostly place, where danger lurked and men were killed as they crawled toward each other's wire entanglements. By day, there was an unearthly stillness, no sign of life, not a bird to be seen. It reeked of the dead.

Occasionally, the stillness was broken by the rattle of machine gun fire which after a few bursts, would quieten down. Or, there was the crack of a rifle as a sniper, ever on his watch, took aim at some reckless fool sticking his head above the parapet and who, more often than not, flopped with a hole between his eyes. Little sympathy did he receive. He asked for it and he got it. Another was the rambunctious guy who, having lived in Canada, and speaking in the vernacular, would shout dire threats and abusive language across the narrow stretch of "No Man's Land." They came back with a volley of bombs which had us scurrying to cover, and then all was peaceful again. It was a game played by both sides.

Other than these marked irritations and the squeaking of the fat, bloated rats swarming in and out of the trenches, gorging themselves on the rotting bodies lying out beyond the wire, the stink from which saturated one's clothes, even the hair and skin. So we sat or crawled into a hole cut in the side of a trench which was misnamed a dugout and tried to sleep.

The war had moved south to the Somme, where all hell let loose and from where the rumble and mutter of thousands of guns could be heard when the wind was right.

Provided one did not stick his head above the parapet, the chances of having breakfast the next morning were pretty good. It was a dull, dreary life, the only compensation of which was a week or so in reserve, a spot of leave in London town, or the indifferent and somewhat dubious pleasure of the joys of a small French town when resting up in preparation for another crack at the enemy. When it would end nobody could forecast, but a person became inured to the monotony, and if he had any sense, he would take things as they were and be happy he was still alive.

As I sat there accepting life as it was, a runner from headquarters came dog-trotting around the corner of a traverse with a field message which he handed to me for my information and initialing. More bilge, I thought. But I was wrong. It was a request from General Headquarters of the Royal Flying Corps (R.F.C.)[1] requesting officers to volunteer for the corps in a flight capacity—pilot or observer. Reading it through again, I thought not of the adventurous life and freedom of a pilot, the hero worship and the extra pay, but merely that it would offer me a few days away from the trenches. If I accepted, the message stated, I would be notified the hour and place to report for an interview. That I would be accepted appeared to me as likely as the war finishing by the end of the week.

Before signing on the dotted line, I walked down the trenches to talk it over with Pearson who commanded a nearby platoon. I gave him the message and after he read it, I said – "How about it? Let's sign on the dotted line. At least we'll get a few days out of this filthy hole – away from the rats, flies, and those stinking bodies out there." But he said that he had no interest in flying and that if we were accepted he would be sunk. I tried hard to convince him but no deal. One reason, he had heard the cavalry was increasing its strength by another division and that the C.M.R. regiments would be mounted again in the near future to prepare for the anticipated break-through at Arras the following spring, 1917.

So, I signed without him and promptly forgot all about it and continued on with the happy and fruitful life of a young lieutenant – swigging rum, which was part of our ration, reading the latest editions of *La Vie Parisienne* and papers from across the channel. As I said before, it was all quiet on the western front except at night when we could hear the horse-drawn vehicles coming down the roads toward the German line. Sometimes, the music of a military band would drift across when the wind was in the right direction and hear the voices of the enemy as they bellowed at the top of their lungs their favorite rathskeller songs. Periodically, we would whiz a volley of Mills bombs across which we pitched like a baseball with an overhand motion.[2] But other than the slight interruptions of machine gun fire, snipers rifles, and hand bombs, the artillery let us alone and it was not a disturbing life. With the heat of summer, the foot-deep mud in the bottom of the trenches dried out and walking became a pleasure. So any dreams I may have had of tearing the skies apart at a terrible 60 mph soon faded away. Little did I realize how this easy way of life would suddenly be brought to an abrupt close.

On August 4, 1916 at 9:30 a.m. I received a field message from regimental headquarters ordering me to report as directed. It read:

To O.C.D. company, 1st C.M.R.'s 4/8/16, Sender's No. B-561.
"Attached hereto is Communication on the genl. Officer commanding the R.F.C. in connection with the application Lt. R. Duncan for transfer to that service. Please hand same to the officer in question and see that he reports as directed."
Lt. Duncan please comply with attached order.
J. D. French, Capt.
O.C.D. Company

At last I'd received the business and was on my way, so I thought, for a few days of easily living in a small French town after being turned down flat. The date was confusing because the British custom is to put the day of the month first – the 4th day of August and not the snappier American custom of August 4th. Anyhow, they had called my bluff and I was on my way. Those few words decided by life's career. Beyond the medical examination of which I had no fear, I felt reasonably sure I would clear the final hurdle and so I said good-bye to my lads in the platoon. I felt rather downcast and guilty leaving them behind, but as a surprise, I gave the sergeant a 120-proof gallon of twenty-year old rum to drink to my health. As I turned the corner of the traverse, I took a last look as those I left behind. I then waved, signaled "drink up," and followed my "batman" who was leading the way to my destiny in the outside world.[3]

We slopped through a long winding communication trench, which was also a drainage ditch, and the one muddy spot in the trench system (it was the middle of a hot, dry summer) to regimental headquarters located in a badly battered chicory factory. Arriving there, I went over to the orderly room to pick up my papers and hoping to duck away without the C.O. Colonel Andros seeing me. I had a great respect for him. A real tough westerner from Manitoba, he was afraid of nothing – Germans, generals, or anything they had in opposition along the Western Front. He wanted me to stay, but realizing I was determined to take a crack at the R.F.C., he wished me luck, adding that if I failed to make the grade, he'd be glad to see me back with the regiment. Again, I felt like a heel leaving the guys I liked so much.

Now for my boon companion, the doctor. I first met him when I was attached for temporary duty to the 28th Battalion. He was also from Western Canada. This was in December, 1915. The Battalion was holding a section of the front immediately forward of Kemmel Hill, a famous landmark in Flanders that gave one an uninterrupted view of the flat Belgium country beneath and far into the distance.

We moved up to the front from rest billets a few days before Christmas and by Christmas Eve were well dug in. The war was practically at a standstill, and other than those on trench duty, everybody was busy opening the many Christmas parcels that poured in almost as heavily as the continuous rains that had hidden the duckboards of the trenches under a foot of mud. The war, to all intents and purposes was over as both sides settled down to celebrate; the liquor flowed freely. Our trenches were extremely close, well within pitching range of those handy baseball-like Mills bombs and we could hear the Germans singing their famous beer hall ditties and sentimental songs so well-known all over the world. A typical German beer hall band was leading them and busting their guts to be heard above the racket.

In the midst of the festivities, the only way to express this incredible situation, the doctor and the quartermaster appeared on the scene primed for anything and ready to go, but unaware of the feeling of good will that was growing fast. Each of them carried a rifle, blasted off a few rounds, and before we could stop them, were over the top and crawling through the wire. Out we went and dragged them back. I steered the doctor to the hole in the wall that was my dug-out, warm and snug from a charcoal brazier. A half dozen other officers crowded in, filling it to capacity, and between the drinks, the warmth of their bodies, and the brazier, the heat became overpowering. Too much for the doctor, he rolled over on his back and went to sleep. The rest of them pulled out and started on a pub crawl down the line which lasted until the early hours of the morning. When we finally turned in, the Heimies across the way were still whooping it up.

When Christmas morning broke, the sun was shining and the skies clear. About noon the Germans began to sing *Silent Night* followed by other sentimental songs. It was really beautiful and made one wonder why we were there to commit mayhem and kill one another. On our side we were still busy guzzling and stuffing ourselves with all the good things that continued to come in an endless stream. So far, not a shot had been fired since yesterday other than the few the doctor and his friend the quartermaster let go. Cautiously we poured over the top of the trenches. Not a shot was fired. Becoming bolder we stood up and looked over "No Man's Land" at the enemy lines which we'd never seen except at night or in the daylight hours through a periscope. Whether this encouraged the Germans is a manner of conjecture, but before long, a huge white flag was waving back and forth, and as we held our fire, a German jumped on top of the trench waving it violently. Others followed suit, shouting and beckoning for us to come over. But this was not the Christmas of 1914 when both sides celebrated as they met and fraternized between the two lines.

Twelve months of trench warfare and with the horrors of the first gas attack a few months before at Ypres where the Canadians held the line without benefit of masks was too fresh in our minds. We were out to kill, not to extend the hand of friendship to men whose leaders had consistently broken all the rules of warfare. We would call the plays, not them. And now we didn't want to play as the whimsy struck them. Yet not a shot was fired and an unspoken truce was declared. The rum flowed freely and from both sides jokes and songs filled the air. The next day we went back to the killing and the maiming with no holds barred.

To return to the doctor, as I opened the door and walked into his office he looked up and said in surprise, "What the hell are you doing here, Dunc?" I replied, "I'm transferring to the R.F.C. and on the way to England to take pilot training just as soon as I've cleared with you." He responded, "A pilot? Why, Dunc, you can't even drive a car so how in hell are you going to fly a plane? Anyhow, sit down and let's get it over with."

He pulled a medical form out of the drawer in his desk, crossed over to a cupboard, and reaching in, came back with a full bottle of Scotch, poured out a couple of drinks, and then we went to work filling in the blank spaces. By the time the bottle was empty and the necessary document filled out, I was in A-1 shape and ready for anything the R.F.C. had to offer. Besides, I had a most pleasant parting with an old and valued friend. He wished me luck and from there I went back to the orderly room where I was given my papers and driven to Army Headquarters which was billeted in a small town a few miles away. I had lunch and then faced the last hurdle – acceptance by an interviewing officer of the R.F.C. And what a pleasant surprise that was!

My lunch had nullified the effects of the liquor and I felt ready for the worst tossing over in my mind of the thousand and one questions that might be thrown at me. By now, just the thought of a few days of riotous living in this small French town had passed and had been replaced with an all-consuming ambition to become a pilot, come what may. I felt quite sure I would be accepted but was afraid that, like so many regimental officers with a background in the trenches, I would first have to complete a tour of duty as an observer – a prospect that had little appeal, and as the politicians say, "One viewed with alarm." My worries were unnecessary.

The interviewing officer was a former cavalry man who transferred to the R.F.C. in early 1915 and with the rapid growth of the corps became a squadron commander within a year. Still in his twenties, he was attached to Army Headquarters and amongst his manifold duties was that of being a one-man selection board for the R.F.C. This he told me as the interview progressed.

Instead of a thorough quiz and cross-examination I was expecting in which knowledge of motors and things mechanical would play a major part, and which I knew would lay me flat, it turned out to be nothing more than swapping yarns about hunting in the Canadian Rockies. The conversation quickly switched in that direction when he discovered that I had been working in the Thompson River Valley in British Columbia with a survey party when war broke out. It was a wild, unsettled area 200 miles from the nearest settlement. By a strange coincidence he knew that section of the country having gone there with a guide hunting bear two years previously.

There being no one else up for an interview, he produced a bottle of Scotch and cigars, then, after a couple of drinks, had my papers made out and transportation arranged for England. I was to report to R.F.C. Headquarters in London for training as a pilot.

Arriving in England early the following morning, I checked in at the Savoy Hotel, and late that afternoon walked over to the "House of Bolo" which the R.F.C. headquarters was so disparagingly called and reported in. The atmosphere was free and easy with none of the stiffness and rigid discipline witnessed at regular army headquarters. Except for a few senior officers, mostly non-combatants, it was a young man's show – teenagers and men in their early twenties. I was escorted to an office by a very beautiful girl – they picked them right in the R.F.C. – where a young captain was sitting, a fellow about my age. He was a former pilot, now grounded due to injuries, and who after he had chatted with me for a few minutes, explained that he was going to post me to a civilian school, the Ruffy-Baumann School of Flying at Hendon Airport.[4] Hendon was a suburb of London only a thirty minute run from Piccadilly Circus by the "underground" as it is called. A billet had been arranged for me close to the field which included a bedroom and sitting room. The house, quite a large one, was occupied by a retired colonel and his wife who opened up their home as a patriotic gesture rather than from necessity. So I had a luxurious billet. London was only thirty minutes away and the next few weeks would be spent enjoying life at the world's most famous airfield. Unbelievable that only two days before I had been a lice-ridden body crawling into a hole in the wall to eat and sleep.

Hendon – The First Five Hours

Learning to fly at Hendon was an exciting way of life, particularly to this carefree youth, full of pep and vinegar, and ready to go. It offered everything. It was the finest equipped airfield in the world, a testing ground for the latest type aircraft and a dozen schools operated by pioneer flyers from Britain, the U.S.A, Italy, and France – all contemporaries of the Wright Brothers. Located on the fringe of the suburb of Hendon, it was an ideal location out in the country. It consisted of a huge field with an excellent turf surface. We didn't have runways then, and was, prior to the war, a rendezvous for racing and exhibition pilots from all over the world.

There the Royal Aero Club[5] of the British Empire checked out those wishing to qualify for the Aviation's Certificate issued by the Federation Aeronautique Internationale (F.A.I.).[6] The Royal Naval Air Service (R.N.A.S.), the air arm of the Navy, had a huge hangar fenced in at one end of the field from which they tested various aircraft submitted for their approval. When I was there in August 1916, they were putting the Handley Page, the world's first multi-engine bomber, powered with two 275 horse-power Rolls-Royce engines, through its final tests. Many other manufacturers also utilized the field for experimental and test flights. It was the hub of every type of aircraft activity in the country.

The various buildings, including hangars housing flying schools, rented space to private owners, as well as sales and repair shops, extended along the village side of the field leaving unobstructed the rest of the area for take-offs and approach. The R.N.A.S. hangar was at the extreme end of the line and anyone stepping inside the enclosure was said to be "coming aboard." At the end of the "deck" and attached to a tall flag-staff was the only wind-cone on the field.

Flying as a sport, prior to 1914, was the prerogative of the wealthy and many of England's noted sportsmen and social leaders took to it with enthusiasm. Indirectly, it was through their efforts and influence that Hendon came into existence and was established to cater to their needs.

Included among the various buildings at the airfield was a lavishly equipped club house sitting in a paddock surrounded by spacious lawns with parking in the back where Rolls Royces and Daimlers predominated. Excellent food was served in the bar and restaurant, and on the lawns when weather permitted. Officers in training were given the "run" of the club and it was there we ate most of our meals and had our drinks. However, the strongest appeal to this flyer's dream was the fact that London was only thirty minutes away by the "underground" which ran twenty-four hours a day.

The following morning, I walked over to the field two blocks away and reported in at the school which was a wooden hangar at the farthest end of a line of similar structures. The office was a small space partitioned off from the rest of the hangar and equipped with a roll-top desk, two chairs, and waste-paper basket. It was there that the owner of the school, Ruffy, was sitting as I entered. Incidentally, the roll-top desk was his pride and joy. It came from the States.

Ruffy was a Frenchman, very friendly and courteous. Slim and short, he was a typical Parisian with a thin face and lively manner which is their hallmark. I liked him immediately. His English was surprisingly good due to the fact he had married an English girl, a former actress and very good looking.

After the essential preliminaries were completed, we went outside where he introduced me to a group of officers who had signed up a few days before, and also to a number of civilian students, still in their teens, who were heading for the R.F.C. once they'd passed the F.A.I. test.

I was more than surprised when I discovered that four of the six officers were Canadians who, like myself, had transferred from their regiments to the R.F.C. This was really something, I thought, starting in with these chaps. They were Jeff O'Brian, Harry Simons, Jamey Stratton, and Dick Hodges, all from Toronto, Ontario, which left me the only westerner. We were the vanguard of hundreds of Canadians who eventually gravitated to the R.F.C. And without the slightest bragging, may I say we gave a good account of ourselves.

There were two French instructors (Ruffy did no instruction work but checked out each student before cutting him loose for solo flying) whose English was none too fluent, yet they were fine instructors and very pleasant fellows. The mechanics consisted of one man who was formerly a Blériot engine and airplane mechanic and a young apprentice. The Blériot[7] was sufficient as an plane, rarely gave any trouble, and in case of a crack-up, had an ample supply of wings and other airplane parts readily available. On the side, Ruffy had a small aircraft factory where he built Caudrons under license – the type we used in the school.

For flight instruction there were two one-seater planes and one two-seater. To keep these operating were two spare engines, an assortment of wings and tail groups, the usual tools and a work bench. Also, several barrels of oil and a good supply of gasoline in two-gallon petrol cans. Gasoline, or petrol pumps were not known. A high-test gas specially distilled for aircraft engines, however, was available.

There was no locker room but an ample supply of nails on the walls to hang one's clothes on, and if one wanted to sit, there were a couple of planks across some barrels that did duty as a lounge. In that wretched English climate as the fall approached, it was a cold and drafty spot. If one "wanted to go," the back of the hangar was the only convenience, but it did offer privacy and was sheltered.

Turned over to one of the instructors, we received our first lesson. An airplane was wheeled out in front of the hangar, the various parts of the plane described and the functions of each. Next we piled into the cockpit of the two-seater, and with the instructor in front and one hand on the joystick, we followed him through as he demonstrated how to warp the wings for turning and the tail plane for climbing. The warped-wing system was the earliest form of control which Caudron continued using long after ailerons and elevators had outmoded them.

Having mastered this, our next step was to learn how to swing the propeller and start the engine. First, wooden chocks were placed in front of the wheels so the plane would not shoot forward as the engine started. That is because brakes on planes were unknown. Then we were shown the stance to take before pulling the prop through, the proper way of doing so, and the various signals used between the man swinging the prop and the aviator at the throttle.

It all seemed rather complicated, and without a proper procedure and maximum care, the danger of being hit and probably killed by the prop would be greatly increased. In fact, many were killed despite the precautions. Four-bladed props predominated and had to be treated with respect.

When the Americans entered the war, many of their air force mechanics were attached to R.F.C. squadrons and they reduced the hazards of prop-swinging by having a man swinging the prop with one hand as another mechanic grabbed on his other hand and pulled him back and away as the prop took hold.

The rotary motors, the 50 horse-power Gnome, 80 horse-power, and the 100 horse-power Mono installed in the Caudrons could not be throttled back but had to be run all out. However, to reduce speed when taxing and gliding, an extra switch was installed on the top of the joystick termed a thumb-switch. By pressing this and then releasing in quick time, the "on and off" procedure would slow up the engine and at the same time keep the cylinders clear and prevent them loading up with gas and oil. This was known as "buzzing" the engine. Later when we became adept with Morse code, we could buzz messages via the thumb-switch. Later models could be throttled back and were used throughout the war allowing the power to be increased from the 50 horse-power Gnome[8] to the 230 horse-power Bentley. Both were also extremely reliable. The crankshaft was fixed like the axle in a car around which the cylinders and prop revolved, creating an appreciable torque.

All three of the planes that we used for practice were 1911 models and extremely light in weight, totally only 600 pounds. But listening to our instructors, we were duly impressed and convinced they were the last word in flying machines. Constructed of wood, wire, and fabric, the nacelle cockpit was placed between the two wings and instead of a fuselage, four tail booms were attached to the wings and supported the empennage (tail plane) and rudders. It was a tractor-type with the engine in front. Other types called "pushers" had the engine in the rear.

The landing gear was an assembly of struts, shock absorbers made of rubber cord, and four bicycle wheels secured to the lower wings.

The instrument board was simplicity itself consisting of a rev-counter, gas gauge, and an oil pulsator – nothing else. Actually, when running up the engine, all we checked was the pulsator, a tube of glass about the size and length of a small thumb in which oil bobbed up and down when it was flowing freely into the engine. There was no warming-up period; we merely ran up the engine, buzzed the switch, taxied out, and took off.

Within two days we'd absorbed all these important factors and were duly impressed by our instructors and the fine qualities of the machines we were going to fly. Now we were ready for our initial flight which was a fifteen-minute flip around the field for the purpose of getting us used to the sensation of flying. It made sense because, for some students, their first flight was also their last. To have the desire to fly is often confused with ability and to force the issue only ends in disaster.

We flipped who would go up first; I lost and so was the last on the list. I watched the other four as they returned to earth smiling and well-pleased with themselves. At last my turn came and I climbed in, with Pierre the instructor in the front seat. He took off and we climbed slowly to an altitude of 1,000 feet, the maximum permitted by the school and then flattened out. I was thrilled to the core. This was it! I knew even before I left the ground that never again would I wallow in the mud of Flanders.

It was a marvelous sensation to feel the thrust of the engine and the ease with which the plane went up and to see the English countryside spread out below on this beautiful summer's day in August 1916 in a checker-board of gardens, fields, and doll houses. The air was smooth and there was no wind to disturb the even motion of the plane as Pierre, with my hand on the joy-stick, had me follow through the first elementary manoeuvers of flight – gentle turns (no banks), wide figures of eight, short glides and climbs. Then we took a couple of turns over the village of Hendon and back to the field for a landing. As I looked over the side I could see the other fellows standing by the Club House.

We were coming down in an easy glide and I was looking over the side watching a soft carpet of green turf as it appeared to rise up and meet us. As the instructor was easing back on the stick to come out of the glide, and preparing to land at approximately twenty-five feet above the ground, a tremendous shape swooped past narrowly missing our top wings and leaving in its wake a turbulent stream of air that turned our fragile little plane, in a flash, upside down and we did a complete somersault and shot out of our seats like stones from a slingshot. I landed on my head and Pierre landed flat on his back. Afraid of fire, I rolled over and away from the plane and scrambled to my feet. Looking around I saw Pierre stretched out, half unconscious, and like a boxer trying to figure out what hit him. I yanked him to his feet and he came to then, brushing himself off. He glanced around and a few hundred feet ahead saw the cause of it all hurrying over in our direction.

As the fellow drew close, Frenchie let go, not in English as he couldn't get that language out fast enough – but in French. The words poured out in a torrent and although I couldn't understand a word he said, it was enough to blister the guy's ears as, at the same time, he flung his arms in the air and literally danced with rage. As this was going on, I walked

Above: Wreckage of Caudron carrying Duncan and flight instructor. Courtesy of R. Duncan Collection at Southern Illinois University at Carbondale.

over to the wreckage where one of the chaps snapped the picture of me.

By the time things had quieted down the source of all the trouble proved to be an American by the name of George Beatty, the owner of the Beatty-Wright Flying School,[9] one of many that operated from the field. He was a great kidder but this time he was hoisted with his own petard. What he had tried to do was to throw a scare into us by flying above and as close to our plane as possible in order to give us the wind-up and render it hard for Frenchie to make any kind of a decent landing. Nothing hurts an instructor's pride more than a rough bouncy exhibition of how-not-to-land.

The plane with which he'd done the damage was a Wright-biplane 1910 model where student and instructor sat side by side with the engine to one side of them. This was accomplished through chains that were connected to two propellers each of which was centered behind the upper and lower left and right wings. Because of this fantastic arrangement, the plane was dubbed the "egg-beater" and looked like it was as well-streamlined as the box cars of the Civil War. It was about twice the size and weight of the Caudron, and although an awkward looking contraption, was on a par with the rest of the planes at the field. It so happened that as soon as war broke out in 1914, an embargo was placed on new models so civilian schools had to be satisfied with what they could get or had before the war began.

Beatty was also under contract to the R.F.C. and was putting a similar group of officers through their paces. He began his fling career in 1908 and for a while was associated with Wilbur Wright according to what he told me, during which time he flew in various competitive events in England and France. Then, having accumulated sufficient capital, he made a deal with Wright and opened up his school at Hendon. As he said, although America was the birthplace of aviation, nobody cared a damn about it over here and the only place where an aviator could make a living was Hendon. We became good friends but after I went to the front I lost contact with him and don't know where he finished up. He had a good line and an endless fund of stories; and he, more than anyone else, was the cause of my eventually coming to the States.

To come back to the wreck, he apologized profusely, paid for the damage and took us both out to the Café Royal for dinner and an evening of pleasure. That was the spot in London much frequented by R.F.C. men and those prominent in the field of art, literature, and the theatre. Men of the caliber of Jacob Epstein, the famous American-British sculptor; Augustus Edwin John, the Welsh painter, draughtsman, and etcher; Somerset Maugham, the English writer known for his plays, novels, and short stories; Elsie Janis, an American actress, singer and songwriter; George Robey, an English comedian, singer and actor; and an assortment of newspaper men and war correspondents.

The contract with the R.F.C. called for five hours of flight instruction which included the flight test for the aviator's certificate issued by the F.A.I., an international flying organization whose objective was to promote flying and aviation in general throughout the world.

I obviously had made an excellent "impression" on my first flight due to Beatty's interference in the landing, but now I was to go it alone as I began my first lesson. This consisted of learning how to "taxi," i.e., practice in moving along the ground from one point to another on a straight line, and keeping the engine sufficiently under control so that the plane would not take off. Both single-seater planes were kept busy as we moved down a designated strip and then turned around and taxied back to where the instructor was standing.

This initial part of our training was completed satisfactorily. We had become expert taxi operators and went into more advanced practice called "straights." Having mastered rudder control and engine operation we were ready for the wild blue sky as it is called in the song of the air force. In doing "straights" we followed the same path as when taxiing only instead opened the throttle wide at the starting point, hurried down the strip at thirty miles per hour, and as soon as we felt the pull on the joystick, eased it back, soared to a height of fifty feet, flattened out, flew level for about a hundred feet, then stuck the nose down and glided to a landing. Like taxiing, we repeated this over and over again until Pierre decided our fledgling hops indicated we were ready to move up into the big time. So, after two hours of tearing up the turf we were ready. Advanced flying was about to begin.

Flight was only possible under ideal weather conditions and with a minimum of wind of not more than five miles per hour. Consequently, the next 1 ½ hours of dual instruction which was split into ten-minute periods, took two weeks to complete. The instructor sat in front, there were no earphones as signals were given by hand or by jerking the joystick. Flight manoeuvers consisted of take-offs, gentle turns (banking), wide figures of eight, gliding, and shooting landings. Maximum altitude was 1,000 feet and we never left the fringes of the field.

Solo was a repetition of the instruction we had received as we prepared ourselves for the F.A.I. Certificate. Tests were given by an official who came down to the field from the headquarters of the Royal Aero Club. Needless to say, we all passed with flying colors. A few days later we received our certificates and were fully qualified to fly anywhere in the world where the F.A.I. was acknowledged.

There was no government authority in those days with rules of procedure controlling aviation activities. The F.A.I. could only suggest and not enforce the simple rules they laid out. It was purely a sporting organization similar in purpose and objective to the A.A.A. of the automobile world.

Under the auspices of Aero Clubs in different countries, instructors were organized and empowered to issue aviators certificates bearing the impress of the F.A.I. It was also these different clubs that organized our meets and races subject to the rules and regulations of the F.A.I. In short, their objective was to promote interest in aviation throughout the world and protect and assist its members in every possible manner. In the British Empire it was represented by the Royal Aero Club who issued my aviator's certificate No. 3702, dated October 17, 1916. Inside was a photograph of the holder stamped with the seal, underscored by the signature of the recipient and counter-signed by the Vice-Chairman and Secretary of the Royal Aero Club. A blue and gold affair, slightly too large to fit inside the average wallet, it accompanied me in my travels abroad, and on occasion, was more useful than my passport. This was not only due to the uniqueness that I had flown in the days of the early birds, but because of the request in the third and fourth pages printed in French, English, German, Spanish, Italian, and Russian which read: "The Civil, Naval and Military Authorities, including the Police, are respectfully requested to aid and assist the holder of this certificate." To be sure, it carried authority and demanded respect.

Before we reported for posting elsewhere, we had an evening on the town with Ruffy who took us to the famous Trocadero restaurant noted for its excellent food and snappy orchestra where he wined and dined his five Canadians, the two instructors, and one American student, a Princeton graduate paying his own way before being appointed to the R.F.C. Hannock was a fine chap and a good sportsman who ran around in a Buick, and whenever he went to town, we piled in and broke all records getting there. Army officers could get away with murder in those days and no cop ever dreamt of writing out a ticket.

Death of a Zeppelin

Hendon also provided one of the greatest thrills of my life, namely, the first Zeppelin shot down in flames as it passed over London and the suburbs. This was the time when the Zeppelins were carrying out a program of night bombing on London. For protection there were the anti-aircraft batteries and searchlights which continually swept the skies hoping to focus a Zep in its beam. Also, the single-seater B.E.'s took to the air immediately when warning was received that the Zeps were crossing the coast on their way to the city. The planes were painted black and armed with a Lewis gun and a supply of incendiary bullets and darted in and out of the beams as they hunted for the enemy. It was a wonderful display of pyrotechnics, and at the sound of the alarm, all of London turned out to watch.[10]

On this eventful evening I was walking from the "underground" to my billet after a night on the town. It was 10:00 a.m. and on my way a raid began over the city. Searchlights were sweeping the skies and anti-aircraft shells were bursting high above and raining shrapnel on the people below. The excitement was tense. Suddenly, in the glare of the searchlights a Zeppelin stood out – a beautiful sight of gleaming silver. Almost at the same moment a sheet of flame shot along its length, the nose dipped and a long blazing object like a comet hurtled toward the earth alight from stem to stern. It crashed a few miles away in the open country, spreading wreckage far and wide across the landscape. As the Zep burst into flames a tremendous roar went up from the millions of people on the streets watching the show, a fantastically weird sort of sound that froze one's blood to hear at such a distance from the city.

Shot down by Lieutenant Leefe Robinson, later Captain, on September 3, 1916, who, as he was cruising along in the dark on Zep patrol, looked over the side of his plane and there to his amazement was a Zep a couple of hundred feet or so right underneath. Down when his nose and from above, in front and alongside, he poured in a stream of incendiaries before it burst into flames. For a moment he thought it would never catch fire and that he would have to tackle the crew in the cabin.

This was the first Zep brought down in the London area and overnight Robinson became a national hero and the pride of the R.F.C. At this time, we were competing with the Royal Naval Air Service (R.N.A.S.) and had beaten them to it! He was awarded the Victoria Cross (VC) the highest award for bravery and decorated by King George V at Buckingham Palace.[11] As a crowning tribute, he also won

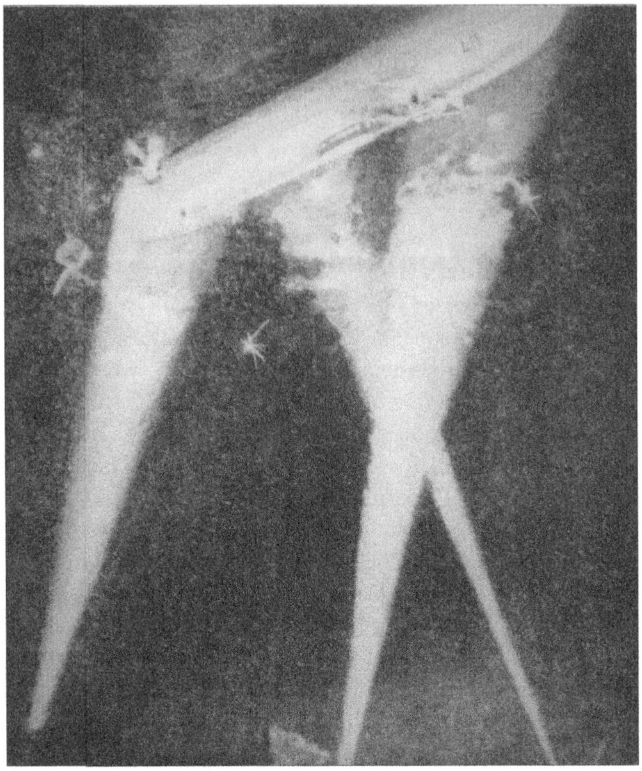

Above: Death of a Zepplin. Courtesy of R. Duncan Collection at Southern Illinois University at Carbondale.

Above: Lieutenant Leefe Robinson in the B.E.2c he used to down airship *SL11*. Courtesy of Aeronaut.

a prize of $50,000 offered by *The Daily Mail* to the first man who brought down a Zep. I met him at the Hendon Club several times along with other members of the squadron. He was a talented pianist and a likeable fellow. Later he went to France with a Bristol Fighter Squadron, was forced down in a scrap and captured. By the following year the defenses of London were further strengthened by a curtain of steel cables suspended from balloons that surrounded the city.

Ground Training: We Go to Oxford

Our next move, after a few days leave, was to Oxford University where the R.F.C. had established a ground school. For quarters they had taken over several of the colleges. Four of our number were sent to Christ Church and I, with another chap, were bedded down at Lincoln.[12] This "morgue" was several hundred years old, and being toward the end of October and cold, the clammy hand of an English winter was seeping in through the doors and cracks of old age. It was no doubt a joy to the antiquarian but a thing of horror to a person used to the comfort of central heating by day and through the night – especially when one has to get up early in the morning. One look at the room and the long stone-floored passage leading to the bathroom and I longed for the warmth and filth of my old dirty dug-out. That, at least, had a charcoal brazier over which one could hover and keep warm.

Some flunkey, I believed they call them scouts in the Army, showed me to the room I was to share with this other chap and which would have delighted the soul of those who liked the quintessence of the 16th century. It was dark, damp, and dismal. The only heat came from a tiny fireplace. The furniture was scanty and at night the darkness was relieved by a small electric bulb hanging down from the center of the ceiling. In the daytime a small window provided us with the minimum of daylight which, like the room, was grimy with dirt. In fact, I'm convinced the light hated to come in and take a look. Such is college life in Oxford. Later I was told a man is supposed to furnish his cell with the normal comforts. The only thing the University provided was plenty of cold water and the Oxford drawl. Anyhow, I called down to Uriah Heap (he was a slimy bastard – all the time dry-washing his hands, leering and reaching out for a tip to pick up my kit and call a taxi. He was rather upset because he had lost a customer and what he could take him for.

I moved over to the Ye Old Mitre Hotel, a public house at 1 Ely Court which dated back a few hundred years but was a fine old establishment, the best in Oxford, warm and snug with many rooms meandering all over the place. My fellow traveler decided to stay with Uriah. He was a rosy cheeked Englishman, brought up on roast beef and Yorkshire pudding so he could take it. A nice lad who made a reputation for himself in the Corps. I was sorry he decided to stay. But he didn't drink and was a rather studious sort of fellow which meant we were much better apart.

The ground school curriculum, which normally preceded flight instruction, covered the airplane and engine theory, navigation, meteorology, and a few hours of shop work – rigging and engine overhaul. However, this was cut down from a month to two weeks as the casualties continued to mount and the supply of pilots diminished. The R.F.C. was scraping the bottom.

Advanced Flight Training

About thirty of us were graduated and sent to various training squadrons for advanced training. I was posted to No. 36 Reserve Training Squadron at Beverley near Hull on the northeast coast of England. In past years, the field was known as a popular race-track. When I arrived, it was bitterly cold with a driving north-eastly wind and heavy rain. Winter was setting in. That worried me little because I knew the tougher the weather, the better the pilot. I had that approach to fitness. This was Western Front weather and the best kind to train in. We had every variety – fog, snow, wind, and rain and the grey cold expanse of the North Sea to fly above. So began my advanced training in November of 1916, and despite weather conditions, we managed to make rapid progress on a series of the more modern types of aircraft.

The Curtiss JH.N. 4D (Jenney) 90 horse-power OX5 engine, U.S.A.
B.E. 2c, 90 horse-power V8, R.A.F. air-cooled engine (Royal Aircraft Factory)
B.E. 12, 140 horse-power V12, R.A.F. air-cooled engine (Royal Aircraft Factory)
Armstrong-Whitworth (Ack -W) 90 horse-power V8, R.A.F. air-cooled engine

The B.E. 12 was a single-seater, the same model as the planes then being used on night-flying for the defense of London against the Zeppelins. The other planes were two-seaters used for dual instruction and solo work. The Ack-W was reserved for night-flying due to its strong landing gear.

Training at Beverley was strenuous. We were kept at it from 9:00 a.m. until whenever, with breaks for meals only, six days a week. On Sunday we studied for the various exams.

Ground instruction consisted of stripping down and assembling the Vickers and Lewis machine guns and correcting jams (cross-feeds); reading and sending in Morse code and artillery observations; covering the code signals; artillery fire; and ranging and corrections.

In air work, we were checked out on all four planes and then cut loose to struggle through alone as we practiced artillery observation, ranging on dummy targets and puff targets (balls of smoke); photography, vertical and oblique; bomb dropping checked by camera obscura; aerial combat (for which we used a gun camera); and formation flying, patrols, radio sending, and cross-country flying. The latter was a flight to an F.E. 2B fighter squadron near Leeds, eighty miles away. On my cross-country, I took off at 10:00 a.m. on a clear day and arrived at my destination two hours later where I had lunch. Leaving around 2:00 p.m., I was only twenty miles from Beverley when a heavy fog began to roll in and with visibility rapidly approaching zero. Because of the danger, I slapped the plane down in the only fair-sized field I was able to pick out. Unfortunately, although I touched ground perfectly, it was a ploughed field whose ruts were frozen and I'd landed across them. These broke the under-carriage and the plane came to an abrupt stop. This was classified as a "forced landing" and not a crack-up, many of which were yet to come.

The field was at the edge of a village nestled around an ancient Norman church and a vicarage almost as old. The vicar insisted I have dinner with him and spend the night at the vicarage, a beautiful old place build in the fourteenth century. Throughout, the walls inside were paneled, the rooms vast and cold, the only heat coming from open fireplaces. Originally it was the Abbot's house of a now ruined monastery. The next day a crew came over, fitted on a new landing gear, and I flew back to the field with one brave mechanic.

This was the last major test except for night flying for which there was no dual instruction. It was too risky and pilots were scarce. The student took up the plane, shot a series of landings under the eye of his instructor standing by the landing strip and then came the test if he was satisfied. Six successful landings were necessary to pass. Few of us failed, but with the little experience, we heard it was rather difficult to judge height in the dark and the tendency was to level off too soon and pancake to a landing. That is where the Ack-W came in with its terrifically sturdy landing gear which permitted the plane to bounce six feet in the air and still stand up under the impact.

There were no field lights of any description on buildings along the boundaries for landing. The only concession was provided by two rows of petrol cans, with the heads cut off and filled with oil-soaked rags which, laid out in an "L" shape, marked the landing strip. The bottom of the "L" was the approach. There were two reasons for this: first, field lights of any description were unknown; and second, the fear of Zeppelins which were crossing over England in bombing raids at night.

Shortly before I left, this complete "black-out" was more than justified when a Zeppelin flew over to Hull, bombed the dock areas and then, at an estimated 4,000 feet, drifted slowly across the aerodrome, turned southeast, and headed back in the direction of Germany. It was not too late in the evening with excellent visibility but the crew never spotted the hangars and other buildings. There they stood, hangars full of planes and with not a machine gun on them. If there had been, by the time their engines were warmed up with guns and ammunition aboard, the Zep would have crossed the North Sea and been safe and sound in its hangar.

The Zeps did a fair amount of damage to the docks and threw such a scare into the population that thousands jammed the main thoroughfares leading to the countryside. All lights were out but visibility was good as the moon was full. Several of us were having dinner on the main street at a famous old restaurant from which we rushed out to watch the show. Suddenly, above the weird sound of thousands of pattering feet, rose the screaming and yelling voices of a mob, ready to tear us apart because we were not in the air. And, no doubt, they would have had we not backed into the restaurant as the inn-keeper locked and barred the heavy oak doors that only a battering ram could break down. So, instead, they turned their vengeance on our tender[13] which stood at the curb, turned it over and over until it was completely wrecked, shouting and cursing the R.F.C. as the pitter-patter of the thousands of feet softened the harsh sounds of fear-crazed men. It was midnight before we finally emerged.

Shortly after, we graduated, sprouted our wings, and with the blessing of the C.O. Major J. A. Cunningham, we packed our kits and headed for London town and R.F.C. Headquarters. On my transfer card it showed a grand total of 22 hours and twenty minutes which included instruction and flying time at Hendon. Two days later we were back at the front, but this time not in the mud. Fifteen of us left together posted to various squadrons, but I was the only one reporting to No. 5 Corps Squadron, the first to land in France in August 1914.

Landing at Boulogne, we boarded a train for Amiens, traveling first-class as all British officers had to in those days. It was going to be a long slow journey so six of us piled together in one compartment to play "21" which in the British Army is known as "black-jack." By the end of the trip, I was a couple of hundred dollars ahead much to the disgust of my brave comrades. Since we were going to different squadrons and they would never have a chance to get back at me, I bought a couple magnums of champagne. We drank them on empty stomachs and then went our separate ways, well-lit and exceedingly happy.

At the depot, a tender was waiting to drive me to the outfit I was to fly with for the next twelve months. The outfit was stationed at Mireaux and operated along a sector of the Somme front slightly north of Albert, a small town in the province of Picardy and famous for the statue of the "Leaning Virgin." The statue hung "unsupported" at a sixty-degree angle from the side of the church tower. I went over to look at it later. Actually it was a lot of baloney, being braced by hidden cables as a morale stunt. Superstition had it that if it fell before the end of the war the Allies would lose. Tricky customers those English. The Russians say they are sly, but I say they are smart. Outside Army Generals, they are always one jump ahead of the other guy.

There was another new pilot going to No. 5 Squadron who had arrived the previous evening and had put up overnight at the Hotel Amiens so we picked him up. He was an Oxford graduate, complete with its distinctive drawl and an air of superiority so many of them affected. Other than that, he was almost human but I had a hell of a time understanding what he was talking about. Oxford men clip their words and leave sentences hanging in the air that only they can grasp. But how he could talk! Mostly about himself and his marvelous aptitude for flying along with his high rating at Central Flying School. This was the original school of the R.F.C. and the training squadrons such as I graduated from were simply part of the rush program without which the R.F.C. would not have been able to carry out the tremendous expansion program just getting underway.

We started out, and before we'd left the center of the city, the truck broke down. That meant we had to wait at least two hours before another one would arrive from the squadron. So we decided to have a few drinks and lunch to fill in time before pulling out. Meeting an Aussie staff captain in one of the bars, he recommended the L'Huite, famous for its roast chicken and soups and where we had one of the finest meals I ever enjoyed at an absurdly cheap price. There was no rationing then and inflation hadn't started.

In the course of conversation, this Oxford birdman learnt how much flying time I had and treated me with a greater condescension than ever. Who wouldn't with fifty-two hours to his credit as against my twenty-two, including that most important twenty minutes. He as a talkative chap who had every move planned and knew exactly how a pilot should behave under every circumstance. Consequently, by the time we arrived at Mireaux, I was convinced that this was the man to hitch my wagon to and follow through the skies as he cleared and freed them from the Huns. Between him and luck, the probabilities of my pulling through the crucial first three months brightened. One thing, however, he didn't know well was navigation. This was my special forte, learnt the hard way in the Canadian bush and the Rocky Mountains and I wanted him to know something of what little useful talent I had, but found it impossible to break through the barrier of words. Perhaps, had he listened, he might be alive today instead of being cut in half by German machine gunners and leaving me all alone to fight my own battle.

No. 5 Squadron

We arrived at the field late that afternoon where I was shown to my quarters in a wooden hut which I shared with another chap. I dumped my kit and then crossed over to the Orderly Room to report. The star pilot was there and the two of us were taken in to the C.O. A short, stocky fellow, and according to our viewpoint, quite old – I guess he was in the middle thirties – he sat behind a desk and scowled at us as we came in, gave us a long-winded harangue on discipline, and then dismissed us with a grunt without shaking hands or wishing us luck. He was a regular army officer of the stuffed monkey school from the Royal Horse Artillery. How he got his wings was a matter of conjecture. He never flew, and I'm happy to say within three weeks he was recalled to England and given a job on the staff. He had guts though and was definitely an efficient officer in the old Army style. However, he didn't know how to handle the high-spirited and reckless young chaps he commanded when they'd done their job and were off duty. Once a pilot came back from the line he was free to come and go around the field. There were no regimental duties to perform as in the Army. Sport, however, was a "must" as fitness was essential to a pilot. For those who wished, there was a stable of six spirited cavalry horses which in a few weeks were made my responsibility due to my being a cavalryman.

The C.O. only appeared in the mess for meals, rarely drank, and talked only to his adjutant and a flight commander who joined him at the table. At the end of the

Above: The Armstrong Whitworth F.K.8 was superior to the R.E.8. Courtesy of Aeronaut.

meal he would stalk out of the room to his quarters which adjoined the mess without looking to the right or left. But he was human deep down – he was in love. She was a beautiful and gifted actress starring in one of the top musical reviews at one of the west end theatres in London who sang a song that made a hit about a lonely girl sitting by a telephone and sang with a breaking heart, waiting for the one she loved to call. Knowing this, whenever we wanted to raze the hell out of him, we would put her record on the Victrola and keep it going until, in a rage, he would stalk into the mess and break it in two. It would always be replaced, and again smashed. Fortunately, before we reached an impasse, he was appointed to the staff at headquarters in London.

There were two other Canadians in the squadron, one of whom was with the 28th Battalion from Saskatchewan when I was attached to them for a short tour of duty in front of Kemmel Hill in 1915 – Lieutenant Jimmy Taylor. He was an observer and a good one but by the time we had got acquainted again, he was posted to England for pilot training. However, in what little time we had together, he taught me much about artillery observation and the ways of the Hun, and how to fly low and live.

For the next three days, Kaylor, the man with all the self-proclaimed honors, and I sat around waiting for the weather to clear. Finally, late one afternoon, the fog lifted a few hundred feet and we were sent up to cruise around the countryside and orientate ourselves with the approaches to the field and the front line. We were given strict orders to keep well behind our line and out of danger from enemy fire from the ground or attack by air. We did just the opposite.

My mechanics wheeled out a dejected looking B.E. 2c and I climbed into the cockpit. As I checked the instruments and fuel gauge, I noticed there was no compass installed or air speed indicator on the instrument board, and no machine gun. I called the flight commander over who was standing by and told him before I took off I wanted those two instruments installed and also a machine gun. We argued back and forth until he finally implied I was scared. That did it, and away I went like a bloody fool. Incidentally, in World War II, he rose to the rank of Air-Marshall. Kaylor, the Oxford fellow who was in "A" flight, pulled out at the same time. How his plane was fixed up I never did find out as the Germans got to him first. But I did notice when in the air he carried a machine gun. Probably because of the passenger he was carrying.

The passenger was an infantry officer who had been attached to the squadron for a short course in liaison, and having completed it, was going on leave the next day. However, before he left, he was anxious to take a flight as he'd never been within sight of the lines from the air. Kaylor was picked for the job because of his far greater number of hours and the impression he already made. He was a convincing talker.

We met in the air for a few minutes, flew together side by side and then, with a wave of the hand and a smile of cheerio, he shot off in a northeasterly direction and quickly disappeared from view as a patch of fog drifted between us and blocked him out. Then it began to thicken – it always does that – and began drifting around on all sides so I decided to cut short the rest of my flight and return to the field. Without further ado, I turned the plane, looked in the direction I figured it would be but saw no sign of it or the woods that sheltered the hangars. It was somewhere in a patch of flog not too far away but in what direction, I didn't know. I had no compass and no idea whether I was flying east, west, north, or south. The shifting fog and the low overcast blotted out the sun or a bearing would have been easy. To add to my predicament, within an hour it would be dark and I would have to come down some place or other before that happened.

It's a very lonely feeling when a pilot is lost under weather conditions such as this with visibility coming and going like a train running in and out of one short tunnel after another and the darkness beginning to close in. But I kept my nose straight ahead, and in the meantime, I picked out a highway lined with poplars on my right similar to that which ran alongside the field. I followed this as best I could as it appeared and disappeared in the shifting fog, hoping it would lead me back home. Then, suddenly leaving a bank of fog underneath my plane, I came into a clear open space over a triangular wood and down below the front lines and the battle scarred ground stood out stark and terrible. I was heading directly into German territory! Almost immediately, I came under heavy fire from an anti-craft battery hidden away in the point of the triangle. The shells were bursting all around, but not too close, and as I looked down, I could see the quick flash of flames from the muzzle of the guns, small but distinct in the gathering dark. What a relief it was to see those battered trees and the fire of those guns outlining the apex because here was the kind of bearing I sought. And for this I had to thank the C.O. who held the theory that an artillery man should know the ground and landmarks in the area within range of his guns. And what was essential for the gunner was equally so for the artillery pilot.

The C.O. had a huge relief map 10' x 20' built up of modeling clay and placed in a special shack for the pilots to study. This fascinated me, having done considerable mapping on survey. Consequently, I spent a good part of my time during the spell of bad weather and previous to my initial flight, studying it. Among the many landmarks outlined thereon, the one that stood out above all others was this small triangular shaped wooded area because of its perfect shape, the point of which indicated, like a natural sign, the direction of the field that lay in a straight line running directly west. For that reason, I swung my plane around, breathed a deep sigh of relief, said to hell with "Archie"[14] who was still throwing them up, and lining up my tail with the Apex, back-tracked my course to the field. I hit it on the nose. It was only about 10 miles away, and I landed by the light of day to the surprise and relief of the squadron C.O. who had given me up for lost.

Although I hated to say so, I admitted I'd been lost as the C.O. came out of his shell and figuratively patted me on the back. But when I told him about the anti-aircraft battery that I had spotted and its location, he could hardly contain himself. For weeks this had been a thorn in the sides of pilots coming within range, and though constantly on the lookout for its location, it had proved impossible to spot until, through sheer luck and as a lost pilot, I had pin-pointed it. At daybreak the next morning two batteries of heavies were turned on the position and obliterated both guns and men. I had chalked up my first kill and it was a good one. It also avenged the death of two brother officers who were literally murdered.

Only twenty minutes after we'd both taken off on that dismal, dreary day, word came back from the front line that a plane had landed in front of the German wire with two men in it, both of whom had been killed. Identification numbers proved it to be ours. The message was jumbled up but it was too late to do anything about it until the next day.

However, early the following morning two of us left for the front line, reported at regimental headquarters where we picked up a runner, and with him as a guide, arrived at that section of the front opposite to where the plane had landed. To stick one's head above the parapet was sure death as a bullet from a keen-eyed sniper with telescopic sights would quickly find its billet. Therefore, we had to look at the scene of the tragedy through a periscope. "No Man's Land" was wider than most stretches, with very few shell-holes but littered with skeletons, remnants of the inglorious Somme show of the previous summer. Where the plane had landed was a smooth soft piece of turf but no a trace of the plane was to be seen beyond a few strips of fabric on the barbed wire fluttering in the breeze, and tire and skid marks left in the spongy ground as it rolled to a stop. According to witnesses, the pilot made a perfect landing, coming to rest parallel to the German wire. As he and his passenger stood up, ready to climb out of the plane, enemy machine gunners literally cut them to pieces and the upper part of the pilot's body toppled over the side of the plane facing our lines. They didn't know what hit them so their agony was short.

On our side of the line, the men were completely helpless. To stick their heads above the trenches and begin firing away would have been a useless expenditure of lives. All that could be done was to watch the proceedings through the periscope and report what had occurred to headquarters.

Throughout the night, the firing never ceased, and under its protection, the Germans opened up a path through their wire, hauled the machine and the bodies back behind their lines and undoubtedly sent them back to the nearest airfield. In the morning, as soon as daylight permitted, the infantry searched the spot with the results as previously mentioned. Also it was photographed from the air, but beyond a well-trampled piece of ground, nothing else showed up in the print. The Germans furnished us with no information as the two men disappeared as if they had never existed.

Endnotes

1 The Royal Flying Corps, or R.F.C., was formed in 1912 and consisted of both a military and naval wing. It had a Central Flying School that provided pilot training for both wings and spent the better part of its first two years testing aircraft for artillery spotting and photographic reconnaissance. Because this type of work, R.F.C. pilots found themselves in aerial battles which caused fighter squadrons to be formed to protect them as they performed their job. As the war continued, the R.F.C. took on other duties such as strafing enemy trenches and bombing. On

April 1, 1918, the R.F.C. and the Royal Naval Air Service (R.N.A.S) were joined to become the Royal Air Force (R.A.F.).

2 The "Mills bomb" was the name given to different types of British hand grenades that had been designed by William Mills. They were the first modern fragmentation grenades and were widely used in both world wars.

3 Batman was the term used by the British Army during World War I for the "soldier-servant" assigned to every officer.

4 In 1915, Italian Felix Ruffy and Swiss Deouard Baumann founded the Ruffy, Arnell and Baumann Aviation Co. Ltd. operated a flying school at the Acton Aerodrome at Hendon. The company became part of the Alliance Aeroplane Company Ltd. in 1918.

5 Founded in 1901 by Frank Hedges Butler, his daughter Vera, and Charles Rolls as the Aero Club of Great Britain, it was given a "Royal" prefix in 1910 and began issuing Aviators Certificates under the auspices of the F.A.I to which it was the UK representative.

6 The Federation Aeronautique Internationale was founded in 1905 by Count Henry de la Vaulx, Vice President of the Aero-Club of France, Major Moedebeck of the German Airship League, and Fernand Jacobs, President of the Aero-Club of Belgium.

7 Blériot Aéronautique was a French aircraft manufacturer founded by Louis Blériot.

8 The French Gnome was an air-cooled rotary engine developed by the Sequin brothers, Louis and Laurent. It was introduced in 1913 by the Gnome Engine Company and was one of the most important designs in aviation history. This engine type became a principal source of aircraft power in the First World War.

9 The Beatty-Wright Flying School was formed by George William Beatty (1887-1955) in February 1914 when he merged with the well-known Handley-Page Company to open a flying school at Hendon. By 1915, the school had five instructors and offered one of the more popular flight training programs for aviators in the First World War.

10 Early versions of the Royal Aircraft Factory B.E. 2 biplane were used as reconnaissance aircraft and as a single-seat night fighter used to destroy German airships between September and December 1916. Its design came under the direction of Mervyn O'Gorman. The term "B.E." stood for "Blériot Experimental."

11 The highest and most prestigious award of the British honors system. The first of these awards were presented by Queen Victoria in 1857 with the investitures held at Buckingham Palace.

12 Lincoln College is one of the constituent institutions of Oxford University located on Turf Street in central Oxford. It was founded in 1427 by Richard Fleming, then the Bishop of Lincoln.

13 A light covered truck.

14 Name given to enemy artillery.

Above: Bristol F2B A7194 and companion in the Middle East. Unusually, A7194 is partially painted white, tried for this hot, sunny climate. Known as the Bristol Fighter, it was an excellent combat aircraft. It was planned to use it as a corps reconnaissance aircraft without its Rolls-Royce V-12, which was in short supply, replaced by a V-8. Courtesy of Aeronaut.

Chapter 2: The Corps Squadron

Above: DeHaviland D.H.4. was used mostly for day bombing and long-range reconnaissance. Courtesy of Aeronaut.

A Corps Squadron was a unit of the R.F.C. attached to an Army Corps for purposes of reconnaissance, artillery direction, and other activities that would assist and also protect it under any and every condition of warfare. An army commander had no direct control over the operations and administration of a Corps Squadron, but was given the fullest cooperation when preparing for offensive action and maximum protection in defense. The tactics employed and the methods of procedure were determined by the Corps Squadron Commander in accordance with those standardized by the R.F.C. The artillery pilot, as Corps Squadron pilots were generally termed, was the most versatile man in the R.F.C. His duties were manifold, demanding that he not only be a skillful pilot, thoroughly versed in the work he had to do, but also a man of stout heart, unafraid and ready to go no matter how strong the odds against him.

In the air, he was invariably out-numbered. For a plane with which to defend himself, and his observer, he was given the slowest and most unwieldly aircraft that the Royal Aircraft Factory ever designed and produced. For the most important of all jobs in the war, on the ground or in the air, the artillery pilot was handicapped with the worst possible weapon to operate and demonstrate that offensive spirit which the R.F.C. had determined was the best defense. While he did pursue the offensive, it was costly in men and machines.

The schedule of the artillery pilot was tough and ambitious. Only thick fog kept the planes on the ground. Drifting fog, rain, snow, sleet, low overcast, thunderstorms, and night operations were taken in their stride. Night bombing, none too popular, was carried out whenever the moon provided enough visibility to line up the targets when objectives deep into enemy territory were bombed. However, with the arrival of the newly organized daylight bombing squadrons in the spring of 1917, the Corps Squadrons were relieved of this onerous duty much to their satisfaction, convinced as they were that the results did not warrant the risk.

The planes, the DeHaviland D.H.4 model, with which these new bombing squadrons were equipped, proved most effective. Able to carry a bomb load of 545 pounds, as against our 80/112 pounds, and fly at an altitude of 20,000 feet – out of reach of most German fighter planes – they bombed almost at will the targets they selected.

Artillery planes performed their duties through a wide range of altitudes determined by the task upon which they were engaged – from fifty feet on contact patrol to 8,000/10,000 feet when on wide-angle photography for battlefield mapping flights. Their activities also included artillery observation and direction, spotting enemy batteries, trench strafing, decoy operations, and on occasion, dropping spies behind the lines and distributing mail over German territory from German prisoners in the British Isles. There was no more interesting job in the R.F.C., nor one that played such an important role in the strategies and tactics of the Army Corps prior to and during major offensives ("pushes") on the Western Front.

Other than photography, when a flight (six planes) flew in formation for mapping purposes only, the artillery pilot worked alone and without an escort for protection unless an exceptionally important task was involved. He flew a solitary course through the skies, a lonely soul with terrific responsibilities while being raked by "Archie" and battered by enemy aircraft. Yet may it be said that these men, who were easy prey to enemy fighters and badly beaten time and again, were never whipped.

Right: Lieutenant R. Duncan with Colleagues. Courtesy of the R. Duncan Collection at Southern Illinois University Carbondale.

Despite their uniforms and dignity of rank, they looked what they were. Barely out of their teens – many still in them – with a sprinkling of ancients in the early twenties. Each and all left their mark in the annals of this First World War. Flying to them was a sport, an outlet, giving vent to the spirit of adventure which was their heritage. They could take a licking and laugh it off, then tuck in their shirts and go back for more with a smile and a wave to the lads on the ground. No grim faced men these, but cheery lads off to take another crack at men with the same spirit, a gallant foe and as game as themselves. It was give and take and may the best man win. Not wanting to die, but unafraid, these gallant lads came from Britain and the lands beyond the seas, including that "rebel colony," the U.S.A.

Every regiment in the British Army was represented in the Corps Squadrons – the Life Guards and Horse Guards, Blue (Household Cavalry), Dragoon Guards, Lancers, Hussars, Artillery, Foot-Guards, Highland Regiments and the Infantry of the Line. All volunteers, most had been blooded and mired in the trenches of Flanders. Others, sportsmen all, and men of wealth and position, came directly to the Corps. Included in their number were many who had qualified for the F.A.I. Aviator's Certificate to insure their acceptance.

There never was a Corps like it and never will be again. These youth, the first in the history of all wars to fly like the eagle and the hawk, took off into the skies, ready to win or die, to meet a foe worthy of their metal. Some lived to blazon the skies with their deeds and become the "aces" of tradition. Others disappeared as quickly as they came. It was exciting as one felt free to move swiftly in space like a bird on the wing. To meet one's foe and win. The rewards were great, not in rank or money, but in the thrills of a fight and the intense joy of accomplishment.

A number of the lads in my flight who were grouped around me, still wore their regimental uniforms and were not required to replace with the R.F.C. outfit until their own was worn out. In the meantime, they carried the wings in their left breast pocket until their regimental badges were replaced with the R.F.C. insignia. A sensible regulation as many returned to their regiments for one reason or another.

Corps Squadrons were not organized until January 1916. Prior to that date, R.F.C. squadrons were a mixed assortment of all types of aircraft, many of pre-war vintage in design, and which operated at one time or another as fighters, reconnaissance, artillery control planes, and bombers.

As the war progressed and developed into the static conditions of trench warfare, the importance of artillery began to be recognized and Corps Squadrons became a major factor in the preparations preceding a "push" and throughout the progress of the battle. This created a need for aircraft specializing in artillery direction and other duties essentially connected with the activities of the troops on the ground. This influenced the High Command, on the advice of General "Boom" Trenchard, Commander of the R.F.C. in the field, to divide the squadrons into two separate Wings, one containing fighter and bomber squadrons attached to the Army, and the other into artillery and reconnaissance squadrons attached to Army Corps. Hence the designation of pilots in the later squadrons as artillery or Corps Squadron pilots.

There were three flights to a squadron and six planes to a flight. Four men were responsible for the operation and maintenance of each plane including the pilot, observer, fitter (engine mechanic), and rigger (aircraft mechanic). Men from other sections of the squadron handled the special equipment determined by the duties in which the plane and flight was involved.

Guns, ammunition, and bombs were the responsibility

of the armorer sergeant-major who kept them in condition and saw to their installation in the plane. Every single round of ammunition was checked before loading into belts and drums to avoid the possibility of "jams." Bomb racks and bombs were attached and fitted under his supervision also.

A photography section was accountable for the care and maintenance of all cameras which were adjusted according to weather conditions. Filters were used extensively in hazy weather which called for expert attention and were outside the province of the pilot. Only when the camera was installed did the pilot take charge to concern himself with the timing between the changing of slides, pushing the trigger ("plunger"), and returning with the pictures he was supposed to take. These were then taken over to the photography hut, and within an hour, the prints would be on their way to headquarters to be used for purposes of building mosaics, mapping, or checking reconnaissance, and artillery reports which they confirmed or refuted.

Radio, or wireless as it was called, was under the supervision of the radio officer ("Sparks") and his crew. In No. 5 Squadron, he was a short, wiry fellow with thick glasses whose greatest handicap was poor eyesight, a sad disappointment which prevented him from earning his "wings." But as the radio officer, he found a thousand and one excuses to get up in the air, ostensibly to check and recheck the radio equipment in the planes. We fell in with his scheming and his hours in jazzing around were almost as many as those spent at the keyboard in his radio hut. The section was on duty round the clock. Their duties also included the installation in the planes of the radio equipment and Klaxon horns,[1] their care and maintenance, and the checking of each radio set as a plane circled the field before heading for the line. This was the last contact with the squadron. Their major repairs consisted of replacing the 200/300 foot aerials which the pilot or observer sometimes neglected to wind-in before landings, and keyboards smashed in combat.

Transport and all supplies were handled by the quartermaster and his staff, often bracketed with the forty thieves, but regardless of such insinuations, they took excellent care of our needs and at no expense to ourselves.

Administration was handled by the C.O. assisted by that great disciplinarian "the adjutant," an intelligence officer to some extent and the flight commander. I served under four C.O.'s in No. 5 Squadron, but Major Eric Tyson, M.C., the last one, was outstanding. He did as much work over the line as the best of us, and more than his share of the lousy jobs. He was killed shortly after I left the squadron for duty with the Home Establishment.[2] Attacked by a combat formation of eight, he put up a great fight, and although mortally wounded, brought his plane down safely. His last words were, "They got me, the bastards – they got me, the bastards."

One of his last letters was written to me in February 22, 1918. I had had a belly-full of England and wanted to get back with him. He was my kind of man. Written by himself on a piece of field message paper this is what he had to say:

22/2/18
Dear Dunc,

I will at last reply. Sorry I have been so long. I have today applied through the Wing for you the end of March and I hope it may come off. I should advise you to shove from that end about a week after getting this. I hope you have considered everything before wishing to come out here again. I don't blame you for it but many people repent of hasty views expressed during a state of temporary insanity. I have heard from Barratt. Williams is out of A.W.'s and I may get him here possibly. Slessor is home and Art an Inf. Coop officer of 28th Wing under Cooper. The Squadron is entirely altered. I now have Colin Tisdall and Phillips as F.C.'s. Poor little Mac went East and has not been heard of. Gilbert also never came back. Willis is knocked out. He was a good conscientious fellow but crashed stunting methinks. The new B.I.O. [Brigade Intelligence Officer] is now finally DePenna, not a bad fellow but no flyer. Russel-Rigby has gone home and been boarded and not allowed to come out again. We have an artillery liaison officer branch in full swing now – i/c Shering. Did you know him? New fellows are coming on fairly well but we have been hit hard and it wants hard pulling to make headway. I have heard from Woodward. Logan is in London with Col. Carthew who is now a big bug I think. No startling news. We have had two fires. One burnt a Nissen and 5 officers' kits, the other last night burnt everything I had and claimed a bit of office bump – Keep fit.

Best wishes and luck.

Yrs. Eric L. Tyson

This was one man I'd follow anywhere. A born leader who never pulled rank except under necessity. He was one of us in the mess and over the line, and was not the only one who tried and didn't get back. His letter, however, portrays the man as he writes, with his best men gone and new ones to train and be blooded – no startling news.

A few days later he went "east" to join the many, all too many, who'd already gone. And I, in the traditional Army way, was sent "west" to Ireland.

The Corps Squadron Planes

From the time the Corps Squadrons were established in January 1916, the activities of the B.E. 2c and the R.E. 8 were restricted to artillery and infantry cooperation. The services they rendered were of incalculable value. Of what they consisted and how they were employed follows the description of these two planes.

The B.E. 2c 90 horse-power R.A.F. engine V8 air-cooled Biplane.

The B.E. 2c was the first plane selected and put into mass production by the R.F.C. for Corps Squadron work. Both engine and plane were designed by the Royal Aircraft Factory (R.A.F.).[3]

Construction was of wood and fabric with wooden struts and wires for bracing purposes. They provided adequate strength within the limitations for which the plane was designed. Pilots who exceeded these didn't live long and no tears were shed when the inevitable happened. The propellers were four-bladed and built of laminated wood of walnut, ash, and mahogany.

The weight and loading was low. Empty, the plane grossed 1,270 pounds and loaded 2,142 pounds. Wing span was 37 feet and length 27 feet. Height 11 feet, 4 inches. Like jockeys, pilots had to be light with an average weight of 156 pounds. Observers also. Maximum speed was 65 mph and cruising 50 mph. Ceiling 10,000 feet which, under favorable conditions, could be reached within an hour. The last three thousand feet was quite a drag especially with a plane that had seen considerable service. Maximum flying time was 3 ½ hours but varied slightly according to weather conditions and the man at the throttle.

Armament consisted of two Lewis guns.[4] The pilot's gun was fitted into a bracket on the right hand side of the fuselage within easy reach of the pilot's right hand. It pointed away from the fuselage at an angle of 45 degrees so clearing the four-bladed propeller by sufficient margin. Line of fire was between the flying and landing wires. The observer's gun was almost as badly restricted as the pilot's due to the peculiar seating arrangement where he sat with his back to the engine and facing the pilot. Between him and the engine was the gas tank which created a double hazard in case of attack. His gun was fitted to a curved racket mount secured to the left side of the fuselage enabling him to move the gun freely up and down but not to the right or left. His job was to protect the plane from enemy aircraft coming in on the tail – the ideal attack when fighting a single-seater but not so good in the case of a two-seater as many learnt to their cost.

Instruments were few but adequate and remarkably efficient for those early days. The compass was equipped with magnets, and any time it went "off," the plane was placed on a "compass rose" – a large concrete surface of packed soil with the points of the compass marked out, and corrections were made with the magnets. Once adjusted and the rest of the instrument functioning, a pilot could fly as true a course, day or night, as the instrument-rich flyers of today. Other instruments included a watch (most efficient), air speed indicator, altimeter, tachometer (rev counter which could also double as a rate of climb indicator), an inclinometer (a bubble in a curved glass tube) for lateral stability, oil gauge, and a gas gauge. The latter was none too accurate so it was implemented with a tank measuring stick.

Accessories consisted of a telegrapher's keyboard and 200/300 feet of aerial wire wound around a drum, a klaxon horn, a Very pistol[5] and flares, and an incendiary stick for burning the aircraft should the pilot be forced to land in enemy territory. Neglect to do this could furnish valuable information to the Germans as was the case with the French pilot Roland Garros who was the first to fly with a front gun firing through the prop installed on his plane. One American pilot burnt his plane alright and then drew his revolver and fought it out with the enemy rather than submit to capture. They had to kill him or themselves be killed.

Under the two lower wings bomb racks were installed each capable of carrying two 20-pound bombs. A larger rack was suspended beneath the fuselage, clear of the under-carriage for carrying a 112-pound bomb. Neither type was carried at the same time.

The B.E. 2c was noted for its stability, easy to handle and a pleasure to fly. But this same quality contributed to its weakness, making it slow on the controls which, combined with lack of speed, rendered it far too vulnerable. It was replaced by the R.E. 8 in August 1917, three years after the first machine landed in France. The B.E. 2c was far from popular, a feeling that was transferred with even greater intensity to its successor the R.E. 8.

R.E. 8: 150 horse-power R.A.F. engine V12 air-cooled. Biplane.

This also was designed by the R.A.F., as was the engine, and put into mass production. A small percentage were built at the factory with the balance of thousands being produced by the larger automobile concerns and subsidiaries. In general appearance it resembled the B.E. 2c with the exception of the upper wings which extended appreciably over the lower. The observer's cockpit had been moved to the rear immediately behind the pilot. A major improvement.

Superior in design and performance to its predecessor, the first series proved far from satisfactory as far as safety and stability were concerned, and it lost favor with mechanics and pilots alike. It was a dangerous machine to fly except in the hands of a skilled pilot and many men were killed until the faults were ironed out – not only on the front line in combat, but during training in England.

The upper wings, with a span of 42 feet 7 inches, extended 5 feet further out than the lowers. Height was 11 feet 4 inches and length 27 feet 10 inches. Weight empty amounted to 1,803 pounds. And loaded 2,678 pounds. Construction was the standard wood and fabric with the bracing common to all biplanes as in the B.E. 2c.

Instruments and equipment were the same as carried by the B.E. 2c but firepower was radically increased. A Vickers front-firing gun[6] was attached to the left outside of the fuselage and two Lewis guns mounted on a revolving Scarff ring[7] for the observer completed the armament.

With a cruising speed of 85 mph and a slow rate of climb, even under the best of conditions, of 300 feet per minute, the plane was an easy prey for enemy fighters and casualties continued to be as heavy as ever.

Structural deficiencies were experienced in the early models. The upper wing extensions folded when subjected to the imposed loads of a steep dive or similar stresses. This was corrected by additional bracing and no further trouble was experienced. But, in the meantime, many pilots lost their lives. Another fault was a strong tendency to spin which was also a cause of many casualties. By increasing the tail-fin area and making several minor adjustments this risk was reduced, but pilots still had to avoid over-control on the rudder and act promptly to avoid stalling from which a spin developed. Once in it, recovery was impossible unless the plane righted itself.

Unlike the B.E. 2c, the R.E. 8 lacked stability and had to be flown with precision at all times. Treated with respect and aware of its shortcomings the average pilot had little to fear. Unfortunately, it was obsolete by the time it got into production. Nevertheless, the Air Board refused to cancel contracts already issued and switch over to other aircraft far superior and available.

Despite poor performance and vulnerability, however, these two planes provided excellent visibility to the artillery pilot and observer who sat high in their seats, head and shoulders above the sides of the cockpit. None too warm in winter, they were protected from the cold with heavy fleece-lined ankle length leather coats, thigh boots, gauntlets, helmet, and fur-lined goggles. In exceptionally cold weather, however, even this protection was hardly sufficient and crews would come down frozen to their very marrow. To combat this, the R.F.C. had a sure bracer called a "Whisky Mac." This was a cocktail consisting of two shots of Scotch whisky and one shot of ginger wine which was tossed down the gullet as speedily as possible, and then chased with a bottle of Bass Ale. The whisky drove out the cold and the ale kept in the warmth. How many "Whisky Macs" depended upon a man's capacity but two usually sufficed. Incidentally, this was also an excellent preventative against the flu that raged the world over in 1918.

Once an army procedure has been put in the book, it is followed to the letter, regardless of time or change. In May, 1918, I was sent to the factory at Coventry to pick up an R.E. 8. This was a year after I'd flown the first one across the line. I reported at the office on arrival and was turned over to a test pilot who took me to the hangar in front of which the plane I was to fly away was sitting. I climbed in, looked over the instrument board, checked the controls, and as I was completing my survey, a procession of four men approached the plane. One carried a bucket of sand, another one a bucket filled with water, the third man carried a fire-extinguisher, and the fourth man was the prop swinger. As they set their paraphernalia down, I looked at them wondering what was coming off and said, "What the hell is all this?" to which one replied, "It's this way, sir. The engine frequently catches fire when starting up and this is in 'case.'" That was a common occurrence with the first series when the rear cylinders would get red hot, but that fault had been corrected months ago. Yet the ceremony continued. They sucked in the gas and, being a sunny warm day, the engine kicked off on the first swing. As I sat warming it up, I glanced over the field and noticed certain sections marked off with red flags. In fact, most of it was. Leaning over the cockpit, I shouted to the sandman, asking him what the red flags were for. "Well, sir," he said, "that's swampy ground." It was a small field and I needed all the runway possible and on looking down its length, I noticed it finished up against the wall of a large shed-like set of buildings. Asking what that was a mechanic replied with a smile on his face, "That's a munitions plant, sir, they manufacture T.N.T." The general effect was that of an obstacle race as I took off, clearing the buildings by a few feet, praying the engine wouldn't "miss," and slowly edging away from the T.N.T and all that it promised. I was in a bit of a sweat till I got clear.

It being one of those afternoons that blossom forth in the spring when the English countryside is at its loveliest, I decided to take my time returning to Andover by flying low and enjoying the beauty of the gardens and woods.

I swung off course in an easterly direction and in a matter of minutes was skirting the edge of Windsor Forest and in the distance as a background, was a tremendous pile of buildings I judged to be Windsor Castle.

There was something I'd been wanting to see from the time I'd landed in England but for one reason or another had never been able to. Now here was my opportunity and I took advantage of it. Determined to explore it thoroughly I came down low, flew across the lawn fronting the royal apartments and in and around the towers and battlements. Then shooting across the Themes I circled the town of Eton College to have a look at my friend Pingo's old school, and back for a final view at that which I'd often dreamed of seeing, never imagining it would be from the air.

Glancing down below, I noticed the streets in the town were crowding up and there appeared to be considerable excitement in the courtyards and grounds of the castle. So I figured it was time to clear out before my engine conked when I would be in the soup. I also noticed the royal standard was flying high which indicated the king was in residence and I began to realize I'd pulled a boner and was vulnerable to a number of charges, including "less majesty."

To put those after my hind off the track, I swung down the Thames for a few miles then, changing course, flew directly to the field at Andover where the squadron to which I'd been recently transferred was located. I was now on the Home Establishment.

The C.O. was Squadron-Commander E.A.B. Rice, M.C., one of the finest chaps I ever met. He was standing on the tarmac as I pulled up in front of the hangar and waiting, as was only too obvious, to have a word with me. I sensed what was coming and I was ready for him. As soon as I climbed out of the cockpit he walked over to be and said, "It took you rather a long time to get here from Coventry. What course did you take?" I replied, "Yes, I know, but I wanted to have a look at Stone Hedge where I landed and wandered around for a while." He laughed, looking very much relieved and said, "I'm glad to hear that because some silly bastard has been strafing hell out of Windsor Castle and the wires are burning up searching for the culprit." To which I replied, "Why didn't they get the numbers off the plane?" Again he laughed, looked at my plane and said, "There were no numbers on it." He guessed it was me alright, but the lack of numbers got us both off the hook as he told me a long time later.

Endnotes

1 The Klaxon horn made an unmistakable sound known as the "Oogah" produced by a motor driven spring-steel diaphragm and rotating cogwheel.

2 R.A.F. Andover was a Royal Flying Corps and Royal Air Force station two miles west of Andover, Hampshire. It opened in August 1917 and remained a part of the R.A.F. until March 1944 when it was transferred to the 9th AF/USAAF. The airfield was returned to the R.A.F. in July 1944 after the American units left for France.

3 The Royal Aircraft Factory was located at Farnborough. It had first been an Army balloon factory but began designing aircraft in 1911. After the formation of the Royal Air Force in 1918, it was renamed the Royal Aircraft Establishment.

4 The Lewis gun was a gas-operated light machine gun designed in the U.S. and mass-produced in Britain. It was used extensively by British and Australian forces as a ground and aerial machine gun. Used .30 caliber ammunition.

5 The "Very pistol" was a signal pistol which derived its name from the inventor, Lieutenant Edward W. Very, of the U.S. Navy. It consisted of a large caliber single-shot pistol with a firing mechanism designed to discharge a flare, blank, or smoke into the air to indicate distress or need of assistance.

6 The Vickers machine gun, originally designed in 1884 by American engineer Hiram Stevens Maxim, was later modernized in Britain and became a standard weapon. It was a water-cooled 7.7 mm and produced by Vickers Limited.

7 The Scarff ring was a type of machine gun mounting developed by Warrant Officer F. W. Scarf of the Admiralty Air Department for use on two-seater aircraft. Due to the weight of the gun, the mount employed a bungee cord suspension that allowed the gunner in an open cockpit to swivel and elevate he machine gun and easily fire.

Left: B.E.2c. First plane selected for mass production by the R.A.F. for Corps Squadron Work. Courtesy of Aeronaut.

Right: R.E.8. Designed and mass produced by the R.A.F. and initially intrended to replace the B.E. 2c. Courtesy of Aeronaut.

Chapter 3: Reconnaissance

Throughout the length and breadth of the Western Front, Corps Squadron planes ploughed through the skies, keeping a continual watch on the enemy along the front and behind the lines from dawn to dusk. Front line work, which was comparatively safe, was concentrated on spotting artillery fire and from the flashes of their guns and locating the position of enemy batteries. Their locations then were marked on special maps for future action, or, if they were raising too much hell and destruction below, we would contact the artillery and in quick time a battery or more would be ranged on the position and wipe them out. This type of work was called an "N.F." patrol. [1]

Batteries were well hidden under camouflage in the ruins of buildings, woods, and semi-dugouts from which the guns would move into the open, fire, and return under cover. They were hard to spot and difficult to destroy. Where good roads were available, it was nothing unusual for batteries to work out in the open. Being able to move freely from point to point, all they had to do should they come under fire was to pull out to another location. However, when a "push" was in preparation, hundreds of batteries eschewed all pretense of camouflage and stood in the open with only a few feet between them.

Unlike N.F. patrols, regular reconnaissance work was tough and dangerous. Pilot and observer had to penetrate deep into enemy territory to collect the information from which the army would determine what their intentions were and be forewarned if preparations for an attack were in progress. We were the eyes of the Army and our purpose was to detect any unusual activities such as the building of light railways used extensively by the enemy for moving heavy guns into position, indications of troop movements, building of new trench systems, and signs of underground diggings for storing supplies and ammunition. With experience and observation, a pilot could read the signs of things about to happen with remarkable accuracy. In short, our job was to detect and forewarn, and then from the many reports received, it was left to Army Headquarters to decide the nest steps.

Our machines were slow, hampered by "Archie" and harassed by enemy fighters – not every time we crossed over the line, but all too often. An easy prey and always outnumbered, many were downed, but there was always another one ready to go and do his best. There were days when it was hopeless to try and cross the line and then we would hang around, cruising up and down waiting for the skies to clear and ready to make the dash should their fighters move in, or some of our own appear far above. It was a case, if one hoped to survive, of the fox outwitting the fox, combined with the patience of Job and everlasting persistence. Sometimes it would be days before we could make the reconnaissance we sought. And then, when we did get across and back with the long awaited report, it became lost and forgotten, or never followed up. This kind of action was far from unusual and it finally contributed to one of the greatest fiascos of the war and a lost opportunity that would have saved thousands of lives and probably shortened it.

I refer to the withdrawal of German troops to the newly constructed Hindenburg Line in February 1917. This was a remarkable piece of strategy in which they moved out of a salient 32 miles in depth on a 25-mile front. This greatly shortened their new position and line, released thirteen divisions for duty elsewhere and was accomplished with comparatively slight losses. Then the Germans sat back in well-constructed and well-drained trenches supported by concrete pill-boxes and backed up with shell-proof strong points. Altogether an impregnable position against the standard frontal attacks of trench warfare.

That the Germans were pulling out had been evident for days, first noticeable by the fires burning in the distance and the black smoke rising to the skies as farms, homes, and villages were destroyed and put to the torch. But there was also plenty of activity in the German trenches as they stepped up the machine gun fire and bomb throwing to cover up their retreat.

Running within a short distance of the trenches was a light railway along which, even in the daytime, they now began moving out equipment and guns. Too deep for us to go in and attack with bombs and machine gun fire but clearly visible to the eye. Enemy aircraft had also appeared in strength in case anyone was foolhardy enough to try.

Then one morning the skies were clear of enemy aircraft, and because activity was noticeable by its absence, we shot across well behind the enemy lines. The first thing we noticed was the disappearance of the light railway, not a tie or nail had been left behind. Nor was there any sign of troops – not a soul was to be seen in the whole of this vast area in which the burning and destruction still continued.

Coming back, we flew up and down the German lines skimming low above the trenches, but beyond a few bursts of machine gun fire there was no sign of life, and as far as we could tell, the trenches were empty. We reported the general situation when we arrived back at the squadron and this time headquarters listened. That night the infantry carried out a series of raids and found no one there. Not even a German chained in his pill box. It was a case for immediate action and the wires burst as calls went out for the cavalry to be rushed up along with light artillery and additional troops. The enemy had disappeared and miles of open territory were free for us to roam around in and nobody knew where anybody was. It was up to the R.F.C. to find out. And we did after frigging around for two weeks or more.

So we took off at daylight the next morning by which time the cavalry had arrived. There was no rain and visibility

Right: Battries standing in open during a "push." Courtesy of the R. Duncan Collection at Southern Illinois University at Carbondale.

Left: Deady crash of pilot and his observer. Courtesy of the R. Duncan Collection at Southern Illinois University at Carbondale.

was good below the overcast. Moving directly east within a few minutes after crossing the line we passed over the burnt out ruins of farm houses and villages, a scorched earth whose desolation extended as far as one could see with miles and miles of open farmlands without a moving thing. Then, in the distance, appeared a checkerboard of lines running in depth to the four points of the compass, straight and true as if on a drawing board. We were gazing upon the famed Hindenburg Line, a network of defenses as impregnable as the Rock of Gibraltar.

When I and my observer, a chap called Craig, had soaked it up and had a clear picture in our minds, I turned and came down low until we were skimming the surface of the rolling ground. It was then, slightly to the front and about two miles to the north, I saw the almost unbelievable sight of a troop of cavalry moving fast in the same direction we were heading. Craig saw them at the same time and I throttled back and shouted we were going in and get set. Then, with my engine wide open, I went after them. They hadn't spotted us, and before they had time to break, we were alongside them and at their own level with both guns going. We shot hell out of them and as I passed, I swung around in a tight bank with my wing tip almost scraping the ground, for another pass but they'd scattered in all directions with a fair number of empty saddles in sight. I was itching to chase them but my mission was urgent and I had to make it back with all speed.

A few minutes later coming in from the west more cavalry appeared. I closed in on them and discovered it to be a patrol of Bengal Lancers, splendid looking men, with beards and turbans, under the command of a white non-com – a corporal.

Although I was in a hurry, I felt compelled to land and advise them of the situation. So I smacked the plane down in a hollow a few yards away from where they were, but deep enough to hide it from sight, then shut off my engine, a very foolish thing to do in cold weather, and waited for them to ride over.

They dismounted and I told the corporal what had

happened and gave him a run-down on what we'd seen, stating I was heading back with the necessary information which would be on its way to army headquarters within a few minutes after I reached the field. Having told my story, I then suggested it would be the wise thing to turn back considering there was no need to push on further and that I would mention contacting him in my report. He looked at me in a most peculiar way and somewhat contemptuously said, "Sir, I have my orders to make contact with the enemy and I intend doing so." "Well," I thought, "you're a bright bastard," and I walked away to my machine and called over, "How about sticking around until I get my engine started?" I thought at any moment some Heinies may come galloping up as they were uncomfortably close. To my amazement, he refused and trotted off with his Bengali turbans and beards. So I gave him the field and all it contained.

My engine was well sheltered and still warm so we had no trouble starting up. Old Craig pulled the prop through a few times to suck in gas, then off it started on the first swing. As I gained altitude I spotted my friend, the corporal, and his dusky men continuing on with their patrol. Deciding he should be taught a lesson I chased after them, and coming in low with Craig firing his machine gun high above them, I almost buzzed their heads off and had those horses stampeding all over the landscape. What happened to them I never tried to find out but I'm quite sure after the jazzing I gave them they called it a day. Without further ado, I steered a true course to the field. It had been a pretty strenuous morning and I was certain my report relative to the cavalry troop would shock the Army into action, particularly the brigade of cavalry. There were miles of open country offering excellent footing ideal for horses and we all anticipated a splendid opportunity of witnessing a brigade of cavalry charging *en masse* like the men of the "Light Brigade."

To my surprise and that of the C.O., the staff expressed doubt as to my report concerning the cavalry troop – their own patrols had nothing to report, and as one gentleman of the old school bluntly put it, he felt I was a bit of a bloody liar or had had one too many. To this day, I can picture myself sitting in the C.O.'s quarters listening as he assured me of his lack of confidence in myself and Craig. Hell, Craig was an old man of 26 who had reached the age of responsibility, a lawyer of integrity who would strongly disapprove of anything in the nature of a faked report. That the C.O. was with us was all important. What the Army thought didn't concern me in the slightest.

Next morning, we were completely vindicated. Four good plugs from the 18th Uhlan regiment had decided that life with the British Army was the better one and had made their way to the brigade horse lines. The C.O. was more than delighted. He'd stuck his neck out for me to a greater extent than I realized at the time and his faith was justified.

Although, our army's strategists and tacticians had lost the chance of a lifetime through lack of initiative and refusal – the reason for which they alone knew – to accept the reports we turned in. Added to which the brash young lads of the R.F.C. became less and less popular with the tough, "blood and guts" old diehards shining bright with spit and polish and entangled almost to strangulation with the rules and regulations of a 1,000 years or more.

The Germans had pulled off a slick move. They'd retired to defenses of fortress-like strength on ground of their own choosing, buttressed with pill boxes and strong points. Under our very eyes they swept their old front bare of everything portable. They shortened their line considerably and had relieved several divisions for duty elsewhere. As far as we knew, they had suffered no casualties in this masterly retirement other than those we shot out of the saddle.

Our score was tragic. We had suffered no casualties worth mentioning but the loss of face to one high ranking officer caused his recall to the Home Establishment. He was a hell of a nice guy, too, but they made him the goat.

The German retirement gave us the opportunity of going through their old lines and comparing them with those we'd occupied at Kemmel Hill and Ypres. So we drove up to see what we could see which was more than we bargained for. A few months before, this had been part of the Somme front. After this petered out with its manslaughter and little gained – 60,000 in one day dead and wounded – both sides dug in, and as usual, the Germans had their choice of positions high enough so that all drainage flowed toward the British lines. "No Man's Land" was once more established but covered with thousands of bodies the British had been unable to move. When we passed through our lines and into the old "No Man's Land," we came upon a most sickening and horrible sight. Hundreds upon hundreds of skeletons in long ragged lines and heaps gradually thinning out until the few who made it part way through the German wire hung there like scarecrows – no clothes not even boots, which the German's had swiped – just bones and weapons. The skulls of many had broken from their bodies as the rats chewed away the flesh. Dentures lay on the ground and the guts inside their vacant cavities were heaped like small pyramids dried and shriveled by sun, rain, and wind. And alongside each little pyramid was a small black heap of maggots that crystallized into the big black flies that tormented man and beast in the summer. Scattered around here and there were wire-cutters. Looking at them, I thought what manner of men are those in high places who send men over the top against a hail of machine gun fire that ripped and tore them apart and with only wire-cutters to hack their way through a barricade of fifty feet of barbed wire. From the badges I saw that many of the fallen had belonged to the same London regiment. I believe it was the Westminster's.[2]

Threading our way through this harvest of skulls and bones, we continued into the German lines. I was astounded at their condition. The trenches were solidly built, dry, with duck boards underfoot, and but for the pill boxes and machine gun nests, it gave one the impression of the narrow streets in a mountain village. Compared to them, ours were

Above: German Gotha G.IV 408/16 twin-engine day bomber. A formation of 21 Gothas shocked the UK by bombing London in broad daylight on June 13, 1917. Two 260 hp Mercedes D.IVa engines provided power. Courtesy of Aeronaut.

Above: German Gotha G.V twin-engine night bomber. A refined development of the G.IV powered by the same engines, it was forced to bomb the UK at night due to greatly improved air defenses. Courtesy of Aeronaut.

hog wallows. As we stood looking around, a chap from the engineers, part of a clean-up squad, walked up and told us not to pick up anything for souvenirs. The foxy Germans had planted them with booby traps completely invisible but sufficiently powerful to blow one to hell. He took us through a dug-out that had been thoroughly screened of which he was as proud as if he'd built it. It was a three storied affair of pit timbers and concrete – the upper was the kitchen, the center was for storing supplies and ammunition, and the bottom story was a dormitory with double tiers of bunks able to accommodate approximately fifty men. Against a frontal attack and without adequate artillery preparation, the position was impregnable as the skeletons above so aptly proved.

All this had a marked effect upon me as I gazed upon the real horrors and actualities of war that was more like Dante's Inferno. The films show them charging and dying gallantly but they do not show the aftermath of rats gorging themselves and the maggots crawling around the inside of their guts. Anyhow, I was more than ever determined to get my quota before they got me and I can truly say I did a pretty good job of "getting," And when I had them cornered I never let up until my ammunition ran out. This I say in no spirit of boastfulness or vanity. The evidence is in the book, in document and picture.

The German retirement – not retreat – was a masterpiece of strategy and organization. Behind the ungodly racket they raised along the front line, they moved a complete army, lock, stock, and barrel several miles to the rear, making no particular effort to disguise what they were about, and then occupied their new and stronger defenses and waited for the British to come up and dig in. Masters of psychology, they knew how the brain of the stereotyped British generals functioned and gambled they would think according to tradition. They did. We lost two planes during this period and neither pilots nor observers were seen or heard of again. One plane I cracked up myself as my engine cut out above the tree tops of the woods at one end of the field. This was my fault. Other than that, it was a cold bitter winter for flying in an open cockpit.

In the early part of March 1917, orders came through. The squadron was to move up north behind the Armentiére front in northern France just short of the Franco-Belgian border. We left Marieux to our regret. Only a short distance away was Amiens, a delightful city and a charming people. Excellent stores, restaurants, bars, and all the good things that bring joy to the heart of a young aviator were there, including my lady love to whom I swore eternal devotion. She and L'Huite, a restaurant famous for its roast chicken and wonderful soups, were hard to leave behind.

Now I'd reached the magic three months and my courage strengthened as I remembered those comforting words after receiving my wings – "If you survive the first three months your chances of coming through will be much brighter. You will be able to fly, to put up a good fight and with luck live on and die happy."

It had been a strenuous three months – getting lost and spotting a long sought after aircraft battery while doing so; crashing into tree-tops; two forced landings without injury or damage to the plane; a cavalry skirmish with 18th Uhlans which I won; being actively engaged in the masterly withdrawal of the German forces to new positions; and a tour of the disbanded German lines. Yet, most gratifying of all was I had learned to fly like a combat pilot and became as one with my plane. Like a cavalryman who sits down in the saddle with no daylight between seat and leather no matter what the gait may be.

Off we flew to Armentiére, home of the famous "Mademoiselle"[3] and renowned for stealing electricity from the Germans. Prior to the war much of their power was piped in from Lyons. After the Germans occupied the city, it was many months before they realized that Armentiére was still enjoying thousands of kilowatts without cost – or to put it more simply, for free!

The airport was impossible, but about the only well-drained patch of land in that area of swamp and desolation, and crisscrossed with drainage ditches and canals. Not only was it small, but the approaches were reminiscent of a steeple chase track. One side of the field was flanked by a canal along which innumerable barges passed daily; on another, hangars and hutments were decided obstacles; and on the north side of the field ran one of Napoleon's strategic highways lined with tall Lombardy poplars. But by the grace of God, the west side was open and clear – the side of the prevailing westerly winds. The B.E. 2c which we still were using, had one good quality – it could get in and out of a very small area. With the R.E. 8 we would have been sunk. One point about flying I may mention is that it remains far easier to squeeze into a small field than get out of it. This I learnt early in my flying career when I smashed through a hedge or clipped the tops of trees.

We had no sooner settled down to the hum-drum existence over a land of swamp and mud, ugly treeless and bare, when the "circus" came to town and we knew the war was won. The Portuguese had arrived. Down the highway skirting the field they came – everything but the elephants – horses, mules and donkeys, goats and sheep – hauling water carts, farm wagons, and army wagons salvaged from the Peninsular Wars or left behind by the Moors. Regiment after regiment passed by, horse-drawn artillery, ammunition and supply columns – all the paraphernalia of an army. They were heading for the battle area and the trenches. From sunny Portugal into the slime and mud of the wettest trenches along the western front that nobody wanted was a stinking trick to pull on them. I'd been through it and I knew the horror they were going into and I felt sorry for them. It wasn't their war but the poor devils were tossed into it for the spoils and politics.

They were skinny, under-sized, wretched looking fellows, mostly, from their appearance, the scouring of the slums with only a sprinkling of healthy, sturdy peasants from the farms and villages. Their uniforms were ill-shaped, bulky, and only fitting where they touched. The officers were slightly better but lacking the air and assurance of one who commands, yet somehow they maintained discipline.

Worst of all was the footwear, varying in shape and appearance from the wooden clogs of the peasant to the pointed high yellow shoes of the continental pimp. Later they were shod with the heavy boots of the British infantryman, the soles of which were studded with rivet-like nails and heels of plated steel.

Altogether, and without exaggeration, they strongly resembled the gypsy tribes of old Russia, or the movie tales of Francisco "Pancho" Villa's peasant army on the march to Mexico City in December 1914 where he met Emiliano Zapata and seemed poised to win control of the entire country. Short of men as we were for front line duty, these poor devils should have been put to work behind the lines as thousands of Chinese coolies were. But the Portuguese government tossed them into the inferno as a token force to fulfill the terms of an ancient treaty made with the British, their oldest ally, after the successful conclusion of the Peninsular Wars.

They were promptly labelled "the Portuguese" by our ribald soldiery who took them under their wing and extended them a certain rough sympathy and hospitality. Learning of the disrespectful attitude of the British Tommy, G.H.Q. issued a cease and desist order the text of which was that, in the future, the habit of calling our brave allies "the Portuguese" must cease. The British Tommies, however, had a rare sense of humor and this really made them laugh. They were wonderful fighting men with happy dispositions and generous hearts. They called everyone "Matey" and never let the lads from sunny Portugal buy a drink. Incidentally, one evening soon after they arrived at the front line, the Germans raided their trenches, hauled a number back to the lines, cross-examined them, took off their pants, smacked them on their backsides and sent them back to where they came from.

No sooner were we settled than we were off again and back in the big time. At Armentiére we were slowly rotting to death. Our new location was the village of Aubigny where, nearby, was a huge field with a squadron of Armstrong Whitworths (Ack-Ws) already occupying part of it. A few miles further west was the fair-sized town of St. Pol which supported a French hotel with a restaurant noted for its food. So we were in luck again. However, there was little comparison between this place and Amiens.

The Battle of Arras was in the preparatory stages in which we were to play a major part. To my delight and surprise we were attached to the Canadian Corps and supported them in their drive for Vimy Ridge. They were part of the 1st Army, responsible for action along the northern section of the Arras front. To their right was the 3rd Army under General Allenby.[4]

It was a happy coincidence. When I left the 1st C.M.R. the previous summer, I had never given them a second thought which was quite usual under the strenuous conditions of war. But I was delighted to know that I flew over them and the rest of the corps and helped pave the way for the biggest battle since the Somme of 1916. General Sir Arthur Currie was the G.O.C., the first real Canadian to command the Canadian Army. His regiment was the Seaforth Highlanders, Vancouver, B.C., of which he was commanding officer when war broke out.

Within the next few weeks our reconnaissance work intensified as we patrolled the front under the most impossible weather conditions. Spotting for enemy batteries, pin-pointing them on our maps, and later forwarding this information through to headquarters kept every pilot in the squadron working hard and against the odds of weather, "Archie," enemy fighter planes, and a none too reliable engine. But it was exciting.

Day after day we pounded the enemy, turning the guns on this position or that, or noting their location for future treatment. Our rugged experiences of the few months preceding were paying off. My observers were good, quick with a gun, and sharp of eye. This was teamwork calling for intense concentration, with a roving eye and a revolving neck as one sought to spot the batteries we were searching for, differentiate between the dummy and the active, and at the same time, keep an eye on the skies for the enemy who now held supremacy of the air and were shooting us down like clay pigeons, and as easily. It was also a period of intense action on the ground.

We flew low, checking the enemy wire later to be cut by our artillery. Wire-cutting by hand was a thing of the past, buried deep in the memories of the Somme along with the bones of the men who tried it. We picked out the pill-boxes, the strong points, the machine-gun nests, and we marked them down, as with eye and camera we strove to reduce the hidden dangers and the hazards against which men had been thrown so recklessly in previous "pushes." We lost men and we lost machines in many a scrap. But men were dispensable in the air as on the ground and those who flew accepted it as such.

On a long distance reconnaissance, we would be given a fighter escort that stayed with us until the job was completed. Their number would depend upon the distance to be flown and the condition of the air relative to enemy aircraft. This was never static as their circus-like formations were constantly moving from one sector to another.[5] There were times, however, when the skies were virtually free and clean and a machine could cross the line with impunity and travel several miles without interference, even in the B.E. 2c, and get back safely.

Intelligence had been hearing from their agents on the other side of the line that the Germans were constructing a second line of defense miles to the rear of their present front. An extension of the New Hindenburg Line, running from Arras south to St. Quentin and beyond. This was in anticipation of the "pushes" at Arras, Vimy Ridge and northwards. Also further to straighten their line and save manpower. Nobody had seen it from the air and it existed solely on rumor. This was prior to the Vimy Ridge show and a few weeks before the arrival of the new fighter planes along with a score of additional squadrons in France. The possibilities of a long distance reconnaissance were discussed and dropped for lack of fighter escort. However, I was convinced that under certain conditions it could be carried out with the minimum of risk given a sky clear of Huns. I talked it over with my observer and he agreed to take a crack at it. He was a young lieutenant from the Gordon Highlanders, Jock Henderson. He had no fear, held the Military Cross,[6] and unlike many airman, favored low flying to the higher altitudes preferred by the Huns. Incidentally, we had not discussed our plans with others in the squadron, not even the C.O., for fear some-go-getter might swipe the idea and add to his credits.

The day came on a beautiful spring morning early in March 1917 – the sun was peeping over the horizon, the skies were clear and dawn was receding fast. We were on the dawn patrol and conditions would never be better. Searching the skies, not a Hun was to be seen. I gave the sign to Jock who sat facing me – we were still flying the old B.E. 2c and it was O.K. with him as he sat tight and ready.

Down we came to the tree-tops, crossed the line without a shot being fired and headed east. In a few minutes we saw our objective in the distance. As we came closer, I climbed higher to get a better perspective, leveling off at 500 feet and there below me and as far as the eye could see north and south lay the new northern section of the Hindenburg Line.

Hundreds of men were working on a complete system of front line support and communication trenches which gave a picture of a huge checkerboard, the lines of which were etched by the light clay from beneath the sub-soil as the slave laborers dug wide and deep. Along light railways, supplies were being hauled as the Germans hurried to complete the work before the well-publicized Battle of Arras got underway.

Another quick survey of this tremendous project and the high sign from Jock who had got the picture also, I pulled the plane round in a vertical bank and into a dive, straightening out about 100 feet above the ground as a protection against attack from underneath. We were getting closer and closer to the line when what I'd been afraid of occurred. Word had been flashed back to the enemy airport at Dotial, and with only about two miles to go, Jock shouted "Look behind, Dunc." I made a half turn, looked up and there were four Albatros closing in fast.

Now it was up to Jock to hold them off as I skidded and slid from one side to another and inched still closer to the ground. This cut out the danger of them coming in on a dive because it left them no room to pull out and the enemy on the ground dared not fire for fear of hitting their own planes. So the bets were pretty even and I felt we'd made it somehow. As the saying goes, "It's all in knowing the angles." It was tricky work with the rudder and I had to work like hell to keep the ship under control. Behind me, the Albatros tried hard to get in those bursts that would bring us down but with an ever shifting target and no room to maneuver. This skidding movement served the dual purpose of presenting a continually shifting target and increasing the arc of fire for my observer. His gun could not swing to the right or left. It was fixed for up and down firing only. Hard to believe, but so it was. For the Huns it was like trying to catch the well-greased pig at a county fair and with a good gunner it was by no means a hopeless situation. I had been caught in similar situations before so I was neither excited or worried as to the outcome.

I watched Jock as he worked his gun and I knew those Huns had a wildcat by the tail as they were forced to sit back and fire at a distance without room to close in. His gun fired over and to the side of my left shoulder and so as he faced me I was able to observe his actions.

As one pan of ammunition was emptied he tossed it into the air, and like a flash, another pan would be in place with hardly a pause in the stream of bullets pouring from the barrel. Although I was keyed up and working like hell as I juggled the plane around, I was fascinated by his skill and coolness to the point where I almost forgot the Huns. But now the bullets were ripping through wings and fuselage and at any moment I was expecting to get it, yet at the same time the line was edging closer and my chances of making it increased. Then with less than two miles to go, Jock held up one finger and I knew he'd got one. He didn't go down but pulled out and back the way he had come. That lessened the odds and made the others more wary. Now there were only three and Jock told me later he had them scared. However, I couldn't see behind me and the only time they were visible was when they came in from the side, passed overhead and like a flash were gone. But my tactics were working fine and every second we were drawing nearer to safety. The tension mounted and Jock's face was a study as he cursed the Germans, shouting and swearing at the top of his voice. Not from fear, but sheer excitement. And no wonder. The bullets were ripping all around the plane and at any minute I expected the gas tank immediately behind Jock to go up or the engine to conk. Those guys were persistent, to say the least.

To gather more speed for the final spurt I stuck the nose down slightly and could see the ground rising slowly toward me. When suddenly the sharp crack of bullets piercing fabric ceased. Scottie eased his hands of the gun, pointed upwards and to the north. I leveled off, looked up into the blue, and turning the machine slightly to the right, I saw the Huns high-tailing it in that direction seeking easier prey at a much higher altitude.

Straightening out the plane, I glanced down below as I headed for home and never did a mass of shell-holes, torn and battered ground, water-filled craters, and those God awful trenches look so good. Also, once a week we had eggs and bacon for breakfast and this was the day. So pulling back on the stick to put plenty of space between us and the mud I climbed to a safer altitude and on to the field. The plane that Jock shot out of the pack we couldn't claim even had he crashed to the ground because we were too far from our lines for any one of our own to check. All claims had to be double-checked.

When we pulled into the hangar, before leaving, we gave the plane a thorough check – the mechanics circling the bullet holes with chalk to be patched later and also checking thoroughly controls, fittings, cables, and engine. In one scrap, I had a rudder cable shot through so that only one thin wire was connecting rudder bar and rudder. In a similar show, one poor devil made it back over the German line and then cracked up on the wire fronting some French reserve trenches.

The report we both worked on was a job every pilot hated but the story had to be told and this was a good one.

Right: German aviator crash into wire fronting French reserve trench. Courtesy of the R. Duncan Collection at Southern Illinois University at Carbondale.

Normally we'd toss it into the orderly room. But this was hot stuff – a first-hand confirmation of rumors that had been seeping through to intelligence and I knew the C.O. would want the dope at once. He was more than enthusiastic when we told him what we seen, and when we had finished, he jumped into the squadron car and on to Army Headquarters and the Wing. The G.O.C. sent his personal congratulations in a field message and later General "Boom" Trenchard who was visiting all the squadrons on a morale boosting tour dropped in at the squadron and said we were a couple of brave men. He was the Commander of the R.F.C. in France.

The Battle of Vimy Ridge in April 1917 was drawing closer but before the troops went over the top, General Currie and the staff of the Canadian Army Corps whose objective this was, were determined that the mistakes of the Somme should not be repeated, the worst of all which was the failure to destroy the tremendously deep barbed wire defenses of the Germans.

The wire cutters had been tossed aside and artillery, ranged and directed by Corps Squadron pilots, began a systematic reduction of those defenses. Impossible to flatten them in their entirety, wide paths at irregular intervals were cut through the wire to give free passage to the troops as they advanced.

Periodic reconnaissances were made to check on the work accomplished or any efforts the enemy may have made to fill the gaps. On these sorties, an oblique camera [See Chapter 8] was carried to be used when visibility was satisfactory to supplement the reports of pilot and observer. This consumed comparatively little time so once the survey was completed, the plane continued on with whatever work had been assigned. With zero hour only a few days away, headquarters wanted another check and I was given the job. Unfortunately, Jock wasn't available and so I took another observer, "Shorty" Carroll, along. As a gunner, he was not so hot, but was an excellent man with a camera, and had a reputation as a newspaper photographer and freelance illustrator.

We were to take off at dawn and so late the previous afternoon we sat down and planned our tactics and other details essential before leaving the ground. At dawn, one's brain doesn't function too well and so I was a strong believer in making all arrangements the day before. Weather, of course, excluded, which no man could forecast in this benighted land of rain, slush, and mud.

This was a job in which there had to be no guesswork or sloppiness as it entailed the lives of many men and I was taking every precaution to insure a good job of work. General Currie placed a far greater value on the lives of his men than most of his contemporaries, and as his staff major told me, he hated the word "expendables."

Periodically we would have a guest night in the mess, a sort of visitors' affair to which outsiders were invited and any officer was permitted to entertain a guest. Dinner was at eight, but long before that the drinks began to flow and everybody loosened up. This particular evening, we had several American pilots over who, in no time, became the life of the party, full of stories and ribald songs. Carroll, who was quite a singer, and a pianist, was quickly taken to their bosom. After dinner the party continued and would until the small hours of the morning. I hated to leave but it was then midnight and it was essential I get some sleep if I was to do a good job. "Shorty," however, was still going strong.

At 4:00 a.m. – the take-off was 4:30 a.m. – I arrived at the hangar, the plane was out in front to go, with the prop ticking over, the engine warmed up, and both mechanics standing by. But there was no "Shorty."

A short distance away was the mess and I could hear the party was still going strong so over I went and there he was – bawling his head off, oblivious to everything but the sound of his voice and the drink in his hand.

He saw me coming in the door and then he remembered. Putting it down and without a stagger he rushed over to me full of apologies. He was lit alright but could carry it. That is one thing expected of British officers. They never brag the next day how drunk they were.

Anyhow, beyond a slight lurch now and then he made it to the hangar where he put on his flying boots, leather coat, helmet, and goggles alright, climbed into the cockpit, checked his guns and camera, etc., and we were ready. The mechanics pulled the wheel chocks and I took off straight ahead. On reaching 1,000 feet I circled the field as he let out the wireless aerial and checked with the wireless shack. Given the O.K. and a couple of more circles to check the engine, I swung over and towards the line. It was almost 5:00 a.m. and dawn was beginning to break.

As we drew close, I looked back at him a couple of times but he appeared to be alright as he grinned at me. Then, for a time, I paid no attention to him but concentrated spotting for enemy batteries the flash from which showed up to advantage in the half light. Too early for Huns to be around and "Archie" also asleep, I cruised along – happy and content – filling in the time before I went low to do the reconnaissance and take the shots. In this manner I whiled away more than an hour spotting and pin-pointing several enemy batteries leaving Carroll to watch for Huns.

Daylight came in full force but the sun was obscured by haze which was fortunate, otherwise we would have been shooting into it. The haze presented no obstacle as the lens were filtered and all indications pointed to us returning with a perfect set of pictures if we weren't shot down by machine gunners from the ground. But we would start in with those guys first and shut them up, or so I thought.

Throttling back to cut down the racket from the engine and prop – the prop makes a hell of a row – I turned as much as possible and shouted to Carroll to get ready, but not a sound. So, half-rising from my seat I looked back and he wasn't there! That gave me quite a jolt and for a second I thought I had tossed him out during my patrol when I was practicing split "S" turns, wing overs, and tight vertical banks. This was the first R.E. 8 in the squadron and I wanted to get the feel of it. However, to make sure, I climbed up another thousand feet, set the tail plane and rudder, throttled back and stood up and turned to look down into his cockpit. There he was, dead to the world, all curled up and with the camera held close to his body. I shouted at him but no response. So I sat down, straightened out the plane and began blurting away with my Klaxon horn. He slept on. Finally, in sheer desperation, I grabbed my Very's pistol, leaned over and, being unable to reach him, threw it at his head. It hit with force but his leather helmet bore the brunt of the blow and he merely shook his head and went on sleeping.

I still had about an hour to go so continued on with my patrol at the same time keeping a sharp eye on the skies for Huns. My luck held and by the time my patrol was up I made it back to the field without incident, without the pictures, and with Carroll still on the floor as useful as the 112-pound sack of ballast we used when flying alone.

I hauled him out with the help of my two mechanics who could hardly contain their laughter and over to his quarters before the C.O. or the adjutant could spot him. There his batman put him to bed.

Carroll woke up about noon, rushed over to the hangar, not having the slightest idea he had been up, bursting with apologies. I said nothing, told the mechanics to keep their mouths shut and went up again with him to finish the job we'd started out to do hours earlier. This time we got into a mess but a flight of Camels came in to the rescue as we battled for our lives. With those guys off our tail, he shot his pictures, swore off liquor until his next drink, and we made many more flights together without further trouble.

Endnotes

1 The term "N.F." referred to hunting for artillery guns that were firing.

2 The Queen's Westminsters was an infantry regiment of the Territorial Army. Designated a Reserve Force, it originated in New Westminster, British Columbia in 1910.

3 The "*Mademoiselle from Armentiéres*" is an English song written by Harry Carlton and Joe Tunbridge and a variation of "*Hinky Dinky Parlez-vous*" and tells of an encounter between an inn-keeper's daughter and two German officers.

4 The Battle of Arras was a British offensive that lasted from April 9 to May 16, 1917. Searching for a breakthrough on the Western Front, it resulted in the furthest advance since the beginning of trench warfare. The British advanced on a broad front between Vimy in the northwest and Bullecourt to the southeast. The Canadian Corps of the 1st Army captured Vimy Ridge. Eventually, it ended in a stalemate for both sides.

5 In 1917 the German Air Army Service created the *Jagdgeschwader* fighter group nicknamed the "flying circus" because of the brightly colored aircraft that they flew.

6 The Military Cross (MC), is a third-level military decoration established by King George V on December 28, 1914, and awarded to captains or officers of lower rank up to warrant officers for acts of exemplary gallantry during active operations.

Chapter 4: The Thunderstorm: The Monsigneur; the Taxi; the Taxi-driver; and the R.E.8

The worst experience I ever had was my battle with a thunderstorm, or rather I should say a combination of thunderstorm and tornado. It was rough and almost fatal. Information had filtered through to Intelligence that the Germans were preparing for a massive counter-attack to regain the ground lost at Vimy Ridge. This came as a complete surprise to us as none of our patrols had noted signs of unusual activity behind the German lines. Neither had there been any of the usual artillery preparations. All was comparatively quiet and peaceful. However, we doubled our patrols and watched for indications.

Then the word came that during the night troops *en masse* had moved up to the front strongly supported by field artillery sufficient in number to lay down a barrage that would flatten and blow our front line to hell – both wire and men. It was early afternoon in the summer of 1917 and according to Intelligence the attack was timed for dawn the next day.

The next day was hot and humid. Over in the east, behind the German lines a storm was brewing, advancing slowly in a westerly direction when an order came through for an immediate reconnaissance of the enemy front and behind the lines. I was given the job, along with a first-class observer, Hill, because of my preference for low altitude flying and experience in such. After we were briefed, we worked fast as the storm was gathering in the distance and we had little time to waste if we wished to beat the weather. The mechanics were warming up the engine as we grabbed our maps, decided the tactics we would use, and headed for the plane. Everything was set, guns and ammunition aboard, radio, and Klaxon horn.

We climbed in the plane. We were now using the R.E. 8 where the observer's cockpit was in the rear, fortunately as it turned out to be. I turned to look back, and Hill giving me the O.K. signal, I began to taxi out for the take-off at the far end of the field. I swung around into the wind which was blowing from the east ahead of the advancing storm. As I sat in the plane checking the motor once more before taking off, the C.O. came rushing over in the car to ask me what I thought about it. By this time the weather had worsened and, in the distance, the sky and horizon as far as one could see had disappeared behind a mass of angry writhing clouds, black as night. There was no chance of getting around it unless it veered which the advance winds increasing in strength indicated it wouldn't. I knew it was touch and go with the odds strongly against me. It was still a few miles back of the German lines but I had about eight miles to go and I was sure that it would be on top of me a few minutes after I reached the front line. So I told him frankly it was pretty hopeless unless the storm veered which was a thousand-to-one chance. But anyhow, I'd take the risk. He told me to sit tight for a minute then drove over to the office and called the wing commander suggesting the flight be cancelled.

Carthew refused – he was the guy I'd had an earlier run-in with. He stated it was of the utmost importance this information be secured. Back came the C.O. to give me the bad news. Then he reached up, shook us both by the hand and wished us luck. As he said later, he never thought we'd make it. Neither did I and it was the only time in my flying career I had the sick feeling in my stomach I'd heard so much about but never experienced. Not exactly fear, but a hopeless feeling of being sent on a job against all odds that could have been held up, say until the storm blew over. But "Tiger Jim" had said the word, so we took off. As we approached the line, I glanced back at Hill who grinned at me and gave a couple of bursts from his machine guns to clear them making sure the bullets sailed into the German lines. We came in low at the south end of the line with my front gun blazing away and Hill doing the reconnoitering with pencil and map and hanging over the side, not missing a thing. In a few minutes we had the story – the trenches were packed with troops and back of the line was a scene of intense activity – troops, artillery, and supplies on the move under cover of the approaching storm. We were the only plane over the line and obviously our visit was unexpected. Then a tap on the shoulder and Hill shouted the O.K. I yanked the plane around to head for home but before I'd flattened out, a terrific up-draft grabbed hold of it and swept us up, completely out of control, into the heart of a black mass of clouds.

For a second or two all was calm. I glanced at my altimeter – it registered just over 4,000 feet. From almost ground level to 4,000 in a matter of seconds. Completely out of control at least it kept right side up and I was still breathing. Then, as quickly as I shot up, I was tossed into the midst of the most violent cross currents tugging and pulling the plane – tossing it in every direction as I struggled with all my might to hold some kind of control. Time and again the stick was almost torn out of my hand or rendered useless. Fighting – not for direction, but to keep the plane right-side up – was all that was possible. And to do that I had to work the throttle as hard as the stick. We were literally flying through water; my goggles filled and I couldn't see. All this time my gallant observer was leaning over my shoulder shouting encouragement. I snatched off my goggles indicating my trouble and he reached further over wiping my eyes and keeping them clear. I cursed the bastard who had got us into

Above: Germans losing ground at Vimy Ridge. Courtesy of the R. Duncan Collection at Southern Illinois University at Carbondale.

this as we plummeted swiftly to the ground like a stone and just as quickly shot up again. Only the engine instruments were working and without one of them I'm afraid we'd have been a couple of dead pigeons. The all-important problem that confronted me was that of keeping the plane on an even keel. Direction meant nothing. I could not even see my wing tips. Suddenly, as like a man inspired, it hit me – the key to the solution was the tachometer. As the plane dived to earth the revs went up as the nose went down, and as we shot back up the revs dropped. Therefore, as the revs went up I pulled the stick back hard and throttled back, and as the revs dropped, I opened wide the throttle and pushed the stick forward trying as I recovered to maintain the normal rpm for cruising. To the initiated it may seem slightly screwy but the combination of throttle and tachometer was the key to the situation and I had to work fast, changing from one movement to another. Instinct and the good Lord were my only guides. In all the various maneuvers I plotted in my spare moments I had never given the slightest consideration to what to do in case I was caught in a thunderstorm. As far as I knew nobody had ever mixed with one and so advice and information was totally unavailable.

Still working away, I expected at any moment to see my wings fold up or the tail section break away. That the cables connecting the control surfaces didn't snap or tear out the fittings under the terrific tension was one thing I began to worry about because this was no severe buffeting like riding out a gale at sea. Instead, I was under attack by tens of thousands of horses-power pulling and tugging at me all at the same time and from all different directions. Visibility was nil, my wing-tips would disappear in the blackness of the clouds and then appear again only to be blotted out as the greys and blacks swirled around us. And so it went on. "My God", I thought. "How in hell will this ship stand up under it?" And then I began to worry about lightning hitting the gas tank as looking down my wings, I saw dancing along the leading edge, tiny lights – a long row of them like fireflies and the same thing on each side of the engine cowling from the front to the rear. Underneath lay a red-hot engine. It was a 12-cylinder V type air-cooled abortion, and back of it was the gas tank and immediately back of the volatile fuel I sat. Bad enough for the plane to fold up but now added to that was the danger of the gas tank exploding if one of those dancing lights popped down the gas tank air vent.

The next thing, the light on my radio keyboard began flashing on and off and at the same time the Klaxon horn in the observer's cockpit started blasting away. For a moment I thought Hill had gone crazy and was calling for help. Turning my head, I shouted for him to cut it out. He grabbed me by the shoulders and shouted in my ear, "What the hell, do you think I'm nuts?" and holding up his hands as the damned horn was still blowing. I thought this can't go on much longer as the battering continued and we were slapped hither and yon. Sweating like a bull, tugging and pulling, a terrific down draft suddenly drove the plane headlong toward the earth then a quick twist and a powerful thrust drove us out from the black and grey mass into a dim barely visible daylight and a sheet of rain as in a waterfall.

The air smoothed out and I reckoned the storm had moved on leaving me behind to my own devices. Within a few minutes I was able to see ground. I was then at about 300 feet. However, my altimeter was still on the burn and I was guessing. Casting around for a spot to land, I saw a large open piece of farm land ahead that offered ample space in which to put the plane down. I pulled back the throttle, went into a glide and was just leveling off when I felt a powerful tug at my landing gear. Down went my nose and deep into the mud of a rain-soaked wheat field sank my prop and engine, and over onto its back in a slow easy motion, the plane. Before we hit, I had snapped off the switches and throttled back as protection against fire. A good pilot does that automatically anyhow. Hill was thrown clear, landing face down in a stretch of mud and water while I was tied up with my legs caught under the engine cowling unable to tug myself clear.

Scrambling to his feet, Hill grabbed the lower part of the fuselage, gave a heave, and I was freed. Looking around to see what I hit I saw a line of telegraph poles and a mess of wires strewn along the ground. As they lay there, hell was popping up and down the front at various army G.H.Q.'s – I'd torn apart the main telegraph lines running north and south. The whole damned war was in a terrible mess and my stock was at an all-time low. All this I learned later. We walked over to the plane, and as I looked down at the wreck, I felt as a man would who had lost a good and faithful horse. It may be slow and an easy mark for the Krauts, but it was built strong as the Rock of Gibraltar. Like a good horse, it had pulled me through. Like "Harry Tate,"[1] it may be comely to look at but it could surely take it. Now the problem was to get back to the squadron before dark.

Right: Devastation of planes and hangers following tornado in Northern France. Courtesy of the R. Duncan Collection at Southern Illinois University at Carbondale.

On the other side of the mess of wires was a main highway, well back from and parallel to the front, and apparently little used by the military. We took off the guns, I unscrewed the watch and put it in my pocket as that is the first thing mechanics swiped in a crack-up, walked over to the road and sat down in the pouring rain to snaffle the first vehicle that came along. It was about 7:00 p.m., according to the watch, and I had to get that report through in a hurry. The quicker it got to G.H.Q. the better.

We sat there, soaked to the skin, puffing on our cigarettes and both aching for a drink and deciding to really have a bender once we hit the mess. When round the bend a French taxi came in sight, chug-chugging away on its two cylinders. I wondered what on earth a taxi was doing in this neck of the woods. We soon found out. I stopped it and Hill, who spoke fluent French, told the taxi-driver we were taking over his cab and he would have to drive us to the nearest Army unit. At that he got very excited when, in the midst of his volley of words, the door in the back opened up and out stepped a very old man dressed in the robes of a Monsigneur of the Catholic Church. It appears that he was returning from the dedication of a rebuilt church in a nearby village destroyed by the Germans in 1914. He was very tired and begged us to permit him to travel on to Doullens as he needed to rest so badly. He looked it, so I decided to let him go on with the understanding that they would stop the first army vehicle or at the nearest unit and explain our predicament. This was an immense relief to him and he was deeply grateful which he showed by having us kneel down on the roadside and giving us his blessing and thanks to God for saving our lives. Then he stepped back into the taxi and drove off. We were rather disappointed as we figured ourselves driving to the squadron in style. Instead, a few minutes later an Army Service Corps car pulled up and we were on our way. The old guy had kept his word and probably thought I would be a convert in the distant future. At the present, however, wine, women, and song were the uppermost thoughts in our minds because tomorrow we might be dead.

We reached the field to meet a picture of devastation. Planes and hangars had been destroyed, hutments blown away, and every conceivable type of wreckage was scattered around. The whole area had been swept by a tornado which left its trail throughout a wide section of Northern France. According to the natives it was the worst storm in living memory. That was the "thunderstorm" into which I had been pushed. Only then did I realize how lucky I was. Out of it came one thing that stood sharp and clear in my mind and that was – given a well-built plane, no matter how rough a time a pilot may have, if he keeps working both man and plane can pull through. There is nothing to fear from lightning except gas leakage. If a pilot has the feel of the plane and the instinct that good airmen have, backed up with sufficient hours and skill, a thunderstorm is nothing worse than a contest between man and the elements. But one contest is enough!!! It is not sure death. There is nothing to fear but fear given an average plane and a pilot who will stay with it.

Those damned Krauts lining the trenches undoubtedly saw me hauled into the clouds and with their queer sense of humor probably laughed their heads off. I often wondered, after my report got through, how many of them were wiped out as the artillery smashed the attack before it started. A show like this is where the artillery pilot evens the score, and when he does, many die.

Surprising as it may seem, I developed a measure of respect for the R.E. 8 – something I never believed possible. Fortunately, I was flying one of the later models which had been strengthened and better stabilized. Also, it was a specially built job, a gift from some town in Australia. Fittings were of silver on the instrument board and other furnishings were superior to the average production job. On one of my leaves, which I usually spent in London town, I met Daphne Pollard and her husband. She was an Australian starring in a West End Review. They entertained me quite often and when I left for the front she gave me a toy kangaroo for luck which I kept in the cockpit. It was with me in the storm. A good luck piece indeed!

The damage was not serious but we could easily have written it off by breaking the long-irons in the fuselage which was the quickest way to rid ourselves of the R.E. 8. But this time I was determined to keep this gallant steed of mine, that battered old hero. The two upper wings had to be replaced, tail-fin and rudder. The engine was O.K. but the prop was broken as were the exhaust pipes with a few dents in the cowling to be straightened. Within a week the plane was ready and rigged to my specifications. It handled like a two-year old and I kept it until I went to the Home Establishment in December 1917. In the meantime, we pulled off many good jobs together and it never let me down.

When I hear people talk about old crates, bird-cages,

Above: D.H.4 bomber with 375 hp Rolls-Royce V-12 Eagle engine was the best single-engine British two-seater.

Above Right: Bristol Fighter (F2.B) with 275 hp Rolls-Royce V-12 Falcon engine. It was the best British two-seat fighter. It was planned to replace the R.E.8 but with a V-8 engine since production of the Rolls-Royce V-12s was insufficient.

Above: S.E.5a fighter with 200 hp V-8 engine. It was probably the best British single-seat fighter and its pilots had a much lower casualty rate than Sopwith Camel pilots.

All photos this page courtesy of Aeronaut.

Above & Right: Front and rear views of the inelegant R.E.8. The 4-bladed propeller driven by its air-cooled V-12 engine are distinctive identification features of the aircraft. It replaced the obsolete B.E. series as the standard British corps reconnaissance two seater. Courtesy Aeronaut.

flying coffins, and the terrible hazards of flying fabric structures, I get rather annoyed. Most of those who talk that way have never flown them, or if they did, it is merely the chatter of the cheap hero striving for a place in the sun.

In my opinion, and former pilots of the R.F.C. would support me, these so-called crates were remarkably sturdy structures. Well-braced and being a combination of wood, fabric, and bracing wires, they had a resilience which gave it the buoyancy of a cork on water. Even the engine bearers (mounts) were of wood (ash) as was the prop—mostly four-bladed. Had the plane been made of all metal construction, it would have been torn apart.

For a short period, our routine reconnaissance work was held in abeyance as the Germans produced a new menace. A much improved twin-engine Gotha.to replace the vulnerable Zeppelins which had been practically wiped out by gunfire and storms. Night raids were out and daylight bombing was in. These began in June 1917 and were so effective that in the following months several fighter squadrons were rushed from the Western Front. Now the artillery pilot was really on the spot losing what little support he had been receiving. This was what the German High Command anticipated. Their purpose was two-fold – to weaken the morale of the British people by ruthless destruction of factories and homes and to force the withdrawal of an effective number of the much-need and all important combat planes with which the R.F.C. had once again regained supremacy of the air. These were the Sopwith Triplanes, S.E. 5's, Bristol Fighters, and D.H.4's for which we had waited so long. They achieved a temporary advantage, but as always, we adjusted ourselves to their strategy by giving them the impression that another "push" was in the making. This was to a major degree placed squarely on the shoulders of the artillery pilot.

We dropped our daily pattern of work and intensified our artillery direction and blasting the front lines, enemy batteries, lines of communication, cross-roads, etc. Trench strafing was a daily task, accompanied by a new and spectacular type of bombing, immediately behind the German lines using magnesium flare bombs along with the standard 20-pound and 112-pound variety. At irregular intervals from dawn to dusk, keeping the Germans guessing and in ignorance of the woefully weakened troops holding the line, we carried out one task after another, around the clock performance where pilots spent most of the daylight hours in the air. For days this continued until reinforcements arrived and we returned to our daily routine. Our losses were comparatively light, considering the many hours flown and the strain upon pilots, men, and equipment.

Endnotes

1 The R.E. 8 was popularly known as the "Harry Tate," a slang reference to a well-known music hall comedian.

Above: Sopwith Pup fighters of the Australian Flying Corps. The Pup was introduced before the Sopwith Triplane. Courtesy Aeronaut.

Above: The Sopwith Camel may be the most famous British fighter. It came after the Sopwith Triplane; it was more robust and had twin machine guns. More Camel pilots were killed in training than in combat. Courtesy Aeronaut.

Left: The F.K.3 was an early British Corps Reconnaissance type. It was superior to the B.E.2 series and was replaced in production by the more powerful F.K.8. Courtesy Aeronaut.

Chapter 5: Search for a Killer

Periodically a special job would come along that required all the knowledge and skill that a pilot had acquired in the hurly-burly of life in the air. One that demanded special qualifications was that of pin-pointing, and accurately, the location of an individual battery. A certain area along the Arras front in the early fall of 1917 had been increasingly harassed by a long-range battery of heavy caliber which was raising hell and destruction on our roads and supply lines. The loss of life and casualties had been heavy. Firing intermittently with a few salvoes at one time, and at another continuing on with a prolonged bombardment, it had the troops nuts to the point where their morale was becoming affected.

Our N.F. patrols had been keeping a sharp lookout but without any luck and so I was detailed to concentrate all my time and effort to bringing the battery to account. My observer, Murray, was an Australian who, being an engineer, applied his technical brain to the problem, and by a series of deductions we narrowed the location of the battery to a small area. From a number of "duds" we learned the caliber of the guns from which we were able to establish the maximum range and from further information relative to the arc made by the hits from shellfire, north and south, we were able to gain an approximate idea within a mile or so of the battery's location. Slow work but the only sensible approach after observer and camera had failed.

Day after day we patrolled the line, concentrating two pair of eyes on every foot of the suspected area waiting for the flash of the guns. But the German battery commander was a wily fellow with the cunning of a fox. Time and again we were driven back by enemy aircraft but only once did they attack.

This was on a sunny afternoon. A few Huns were flying around in the distance too far away to give us any concern. But we kept our eyes on them as we combed the ground below for this most elusive of batteries which still had us guessing. Suddenly, there was the sharp crack of bullets as they pierced the canvas, a sound that no engine can drown. Automatically I yanked the plane up and into a climbing turn and saw to my consternation, five Albatros coming in at us. The man in the lead with both guns firing, headed straight for my tail. As he passed on down, the others followed suit like hounds after a rabbit. Murray was doing a fine job holding them off as I twisted and turned trying to keep their aim off the plane and give him a clear field. He winged one and then another who pulled out but didn't go down. This left us with three who had a wildcat by the tail and didn't know what to do about it.

But instead of changing their tactics, in they came again in the same old dive which clearly indicated they were new at the game. All three shot past and then it was quiet and I thought we'd beaten them off when a burst of fire from underneath the cockpit smashed into the instrument board, fortunately missing the fuel gauge and old pressure lines and almost got me. One bullet grazed the inside of my calf from heel to knee as it ripped through my flying boot, the scar of which I carry to this day. A little further to the right would have been the end and also death to my observer as we crashed below. However, one was a real sportsman, who making an "Immelman climbing turn" shot past us and almost grazing our wing-tips as he headed east waving his left hand and smiling all over his face. He was the second German pilot to do that after a damned good scrap. I couldn't give him the high sign but Murray did as he held his fire and cheered him on.

The friendly interlude over, we continued with our hunting determined to stick with it until we got our man. Several days later our resolutions were beginning to weaken when we spotted them with all four guns blasting away at a battery of heavies stationed about three miles behind our lines. They were right on target but luckily our gunners had taken shelter. We were then at an altitude of about 2,000 feet and the battery stood out clearly in the open. First pin-pointing their position on our maps, I throttled back and began gliding down slowly in a wide circle at the same time shouted to Murray I was going in and to grab his camera and snap them from the rear. We shot over on the right about 100 feet with old Murray clicking away like a press photographer and were gone and out of sight before they realized we had seen them. He did a good job as we then flashed their position to a counter-battery that were on the alert. Two batteries of heavies went into action like lightning and put them out of business; by dark they were silenced. Only once did we have to re-range the guns.

The following day the site was photographed and the print showed the barrels of the four guns were blown clean off their mounts but still intact, a huge crater marking the spot previously used as an ammunition dump and the surrounding area was pock-marked with shell holes. The crew undoubtedly headed for their dugout, or fled if one wasn't available, as no bodies were visible. Being alongside a good road they could move fast.

Looking at the pre-shoot picture it can be readily seen how the law of the land worked in their favor. They were in a hollow, the guns blended into the landscape, and being in the open with a good hard road nearby, they were exceptionally mobile and could be moved from one site to another with comparative ease. However, we caught them short as within minutes after we returned to the field another plane was up to find the gunners were still pounding hell out of them. With the exception of the largest caliber, guns were horse-drawn and it took time to harness up and get underway. So even if they had realized in time, they were spotted and doomed.

Right: The DFW C.V, powered by the 200 hp Benz Bz.IV, was built in greater numbers than any other German warplane. This example was built under license by the large LVG company. Not especially fast, the DFW C.V was robust, maneuverable, and had good handling qualities. It served at the front from late 1916 until the Armistice. Courtesy Aeronaut.

For this show and a combination of other jobs we had pulled off we received the congratulations of the Wing Commander Colonel "Tiger Jim" Carthew and General Trenchard who hadn't forgotten me, which resulted in us both being awarded the Military Cross, July 29, 1917. This show was also my first experience with the "Immelmann turn" and only luck saved me from disaster – and my mascot, the little Kangaroo. This was credited to Immelmann the great German ace. Adapted by our fighter planes, it proved most effective. Fortunately for many of us, Immelmann met his death in the summer of 1916 (June 18th), shot down by a nineteen-year-old boy whose total time in the air was only four months – Flight Lieutenant McGubbin. It marked the end of a Fokker Eindekker II and a pilot to be reckoned with.

One artillery pilot flying a "Quirk" as the B.E. 2c was dubbed, described his first experience with the "Immelmann turn" maneuver by saying "The God-damned plane was going and coming at me all at the same time." Here's how it's done. From level flights, dive the plane slightly with throttle wide open to gain maximum speed, then pull up into the first half of a loop. When upside down push the joystick forward and to one corner of the cockpit and at the same time apply full opposite rudder. As the plane rolls around toward an even keel, pull the control stick back and directly across to the opposite corner and swing on full opposite rudder to stop further rotation. The plane will now be headed in the opposite direction so straighten out the rudder and, as the plane levels off, push the stick forward until speed has been regained. Snap and precision on the controls are the main essentials.

Those slow artillery planes, defenseless as they were against the mass attacks of the enemy, could give a good account of themselves in the hands of a skillful and resourceful pilot or none would have lived to tell the tale. Numbers could also be made to work to their advantage by going straight into the formation and mixing it. This, according to German thinking, was not playing the game and got them confused and in the consequent mix-up and crossfire they were subject to the danger of bringing one another down. The greater danger, however, was that of collision which occurred more than once.

Used to seeing so many fledgling pilots turn tail and dive in a futile effort to get away, there was sound reasoning in reversing this normal procedure but it takes terrific control in overcoming the natural tendency to turn tail and run in the face of overwhelming odds. Five or six is not too bad but as these formations grew in size, some reaching 30/40 which was overwhelming. But to attack was the only hope. Furthermore, the probability of our own fighter planes being around and swinging into the rescue turned the tide in my favor more than once and allowed me to make my getaway which in a dogfight is none too difficult.

Bluffing also paid off. In one scrap where the odds were dead against me, I threw the machine supposedly out of control with full engine, slumped in my seat, my observer flopped to the floor, and to all appearance the plane was twisting and turning on its way to the ground. Actually, although slumped in my seat, I was working the controls and throwing the plane into a series of right and left hand turns – a maneuver described as the "Falling Leaf." My version was somewhat ragged but it kept the plane from going into a spin and deceived the enemy.

I practiced continually on my routine patrols and without fear of the plane coming apart. The rough experience in the thunderstorm was ample proof the plane could take it since it was re-strengthened. But spinning had to be guarded against as it was virtually impossible to bring it out of one. Many deaths resulted which included test pilots and many others with ample flying time among the victims.

Unfortunately, in the hurry to push men through their training at maximum speed, there was no time to discuss with those who were being trained for reconnaissance and artillery work, factors of safety and the complex features of aerial maneuvers. As long as a pilot could bank and turn, climb and glide, and steer a straight course, direct a shoot, and carry out the manifold duties of a Corps Squadron pilot, he was ready for the front. As General Trenchard repeatedly stated – "The only way to learn how to fight in the air is by doing it. There is no time to waste on frills."

He was right, but hundreds of pilots lost their lives who would have lived had they received a little more training in the "tricks of the trade." Some on their first mission, and many long before the crucial first three months was completed. The reconnaissance pilot certainly had it rough.

German Corps Squadrons

The L.V.G. powered with the 130 horse-power Mercedes and later the 175 horse-power engine was the first of Germany's reconnaissance and artillery machines exclusively devoted to this class of work and also daylight bombing. Later the Aviatik C II and the Albatros C III powered with the same reliable Mercedes engines of 160/170 horse-power carried the major load of the artillery and reconnaissance activities of the German Air Force.

In all-round performance, they were superior to the B.E. 2c and the R.E. 8 and were particularly fortunate in being powered with the Mercedes engine whose infallibility was only exceeded by that airman's delight, the Rolls-Royce motor. Maintaining the policy of the German Air Force, they remained strictly on the defensive and rarely crossed our lines and so their losses were considerably below those of our Corps Squadron planes. However, during periods of R.F.C. air supremacy they were shot down with impunity even though they had the advantage of a covering screen of fighter planes.

This cautious attitude was not conducive to accuracy in ranging artillery, the reliability of reconnaissance reports, or aerial photography. The whole policy of the Germans, both in the air and on the ground, was a defensive one in the belief that attrition through the terrible loss of the Allied armies would eventually swing the fortunes of war in their favor. In the air it had certain merit. It meant that their planes, if brought down, the injuries to the crew would be slight and they would be returned to service.

Right: LVG C.II 342/16 in flight. The C.II was powered by the 150 hp Benz Bz.III or 160 hp Mercedes D.III. Not a brilliant performer, it was robust and easy to fly. Courtesy Aeronaut.

Left: LVG C.V was powered by the 200 hp Benz Bz.IV. Designed by the same designer of the DFW C.V, it had a much better performance than the earlier C.II. Like the DFW it was robust, maneuverable, and easy to fly. One of the better German short-range two-seaters, it was produced and used in quantity. Courtesy Aeronaut.

Chapter 6: The Decoy Plane

The German trench systems, which were actually in-depth fortifications, included reserve trenches and other strong points that paralleled the front line and were remarkable examples of field engineering. In comparison, the British lines could be classified as nothing else but open ditches, some better than others, riveted with sandbags which were continually replaced as they rotted. Water-logged and ankle deep in mud, except in summer, conditions were only relieved by duck boards in isolated spots. They were accepted as only temporary, a fallacy which we were encouraged to believe to convince us that in the next "push" we would have the Germans on the run with their final stop Berlin. Such was the power of propaganda during the years of 1914/18. When the war settled down into the stalemate of the trenches, the Germans were able to select their lines of defense and take advantage of whatever stretches of high ground there were while the vast lowlands of Flanders quickly reverted to the swamps of ages ago as the drainage ditches and canals were destroyed.

Along with the added advantage of an unlimited supply of slave labor, and a more realistic appreciation of the duration of the war and the course it would follow, the Germans sat back behind the protection of what was a veritable fortress stretching from the northern coast of Belgium to the borders of Switzerland. At irregular intervals, and in every sector, this front line was strengthened with solid concrete structures termed "pill boxes," protection for machine gunners able to fire in all directions except directly overhead. Coupled with these were machine gun nests and sand bag enclosures from which hand-picked machine gunners were able to fire in every direction as well as overhead. These were a thorn in the side of attacking troops and were also held in respect by low flying aircraft whose vulnerability was demonstrated by the number shot down. Including among them, Major Edward "Mick" Mannock, England's leading ace with 61 victories.

To put the many machine gun nests out of business was an impossibility due to their mobility and so this never became a regular routine job for the artillery pilot. A menace at all times, and particularly at night, they swept "No Man's Land" with intermittent machine gunfire seeking out our men as they carried on with their night-time activities in anticipation of a "push." The lessons taught us at the Somme in the summer of 1916 dictated how important it was that the effectiveness of pill-box and machine gun nests should be neutralized to the lowest possible factor. The Battle of Arras was pending. Various schemes had been tried to pinpoint the exact location of these key positions by having scouts sent out at night who crawled along or beneath the enemy wire and moving from point to point in the direction of heavy concentrations of fire-power. But they had to return before dawn and it was a slow tedious job that bore little fruit beyond guesswork. Photography was a failure although we tried both vertical, oblique, and stereo shots. Their positions blended too well with the trenches and whatever camouflage they resorted to. So the decoy operation was devised, the success of which demanded a pilot skilled and able in handling a plane at low levels, an observer with plenty of guts, a sharp eye, and a plane with an engine that would not cork and was reasonably responsive to the controls. I had both the observer and the fastest R.E. 8 on the Western Front. One that handled like a fighter plane. So I got the job. A plane rigged according to my specifications meant it was tight and quick to answer the stick.

The infantry with whom we were cooperating provided the ground contacts. We were fortunate inasmuch as this had been tried previously although still somewhat of an innovation. Anyhow, we were able to profit by the mistakes of others. The trial and error system was the only way we learnt in this first war in the air.

The procedure was comparatively simple. Men of above

Right: German soldiers escaping bombardment. Courtesy of the R. Duncan Collection at Southern Illinois University at Carbondale.

Above: Remains of burning blane and charred body inside. Courtesy of the R. Duncan Collection at Southern Illinois University at Carbondale.

average intelligence with blackened faces and camouflaged clothing were selected to crawl up and under the German wire before dawn broke, well-hidden from the eyes of any German foolish enough to stick his head above the parapet, and in the neighborhood of where they considered their objectives were entrenched. Our job was to fly up and down and immediately in front of the German line and so draw the fire of our disgruntled foes who normally would be sleeping away in the comfort of their deep dugouts. We knew that for any unusual manoeuver the call for arms would sound and they'd line the trenches *en masse*. I picked "Had" Driscoll to fly with me, a Canadian, Princeton graduate and a professional pianist – one of the best in the concert world and also one of the finest observers I ever worked with. Fearless, he was tops at his job and with a decided hatred of anything smelling of sour-kraut and garlic.

We took off on a fine, windless day with a slight overcast and not a bump in the sky. A perfect day for the job, with smooth air right to the ground, easy on the engine, the plane, and the pilot. We were all set. "This was Had's show and I'd turned everything over to him, from the planning with a staff officer at corps headquarters, to spotting and pinpointing the positions we were after. Before he left headquarters, I gave him a small carton containing Mills bombs with the hope and request that he bounce them off the heads of those big-mouthed Heinies who were still shouting violence and mayhem across "No Man's Land." He did.

Dawn was beginning to break as we arrived at the line, and shooting up a couple of red and green flares to alert the scouts, we came in from the flanks with streamers flying, not above the German line but parallel and at sufficient distance to permit the men in the pill boxes to line their sights on the plane as we traveled across their front. We looked like an easy target, but it was far from such as I skidded rapidly in irregular flight down the course. The plane handled beautifully, the engine purring away, the rigging taut and tight, and with the controls quick to answer the slightest pressure of stick and rudder. I sat tight, like a man in the saddle, with a spirited animal between my legs and wished that my mechanics, Shearer and Little Tick, were riding in the back to see and feel the results of their handiwork—the miracle they had accomplished.

I thought about them later as I continued the patrol with the appreciation that existed between pilot and mechanic and suddenly awakening to the peculiar circumstances surrounding the plane. It was built in England, designed by the most God-awful engineers in the world, was a gift from a city in Australia, flown and operated by two Canadians, and serviced and virtually reconstructed by two English mechanics – an unbeatable combination which included my miniature kangaroo.

The first run caught them napping as I slid in from the flank and drew little fire, but on swinging around for the second run, I was met with a hail of fire and the sharp crack of bullets piercing fabric. Once more we came in, still holding our fire, but further back from our line. From there we made a wide swing round the northern flank at a higher altitude that would offer sufficient gliding distance to a spot behind our lines in case we were hit in a vital spot, or one or the other was wounded, and made the final survey from behind the enemy lines which gave "Had" the opportunity to size up the situation from another angle and with the added possibility of drawing fire from strong points located back of their front. The German strategy was based on defense in depth, not the thin "red-line" of mud that we relied upon. On the face of it, this looked more hazardous than our previous tactics but actually such was not so due to the element of surprise. In short, we got them off guard and confused them with the unusual.

Throughout we had not fired a shot but now were ready for action with front gun, two rear guns, an extra supply of ammunition and the friendly little gifts from the staff at headquarters. So I pulled back across our lines and out of range to get my second breath and allow "Had" to complete his notes and loosen up his guns. Then back we went. This trip I flew directly over their trenches with nose slightly down so my front gun could rake them better and with "Had" giving it to them from the rear. The pill-boxes covered with concrete were harmless but the machine gun nests gave us the business, and as I passed over, I could see the gunners' faces distinctly as they swiveled their guns around and up. The flight continued until our ammunition was spent and we had let go the bombs. By this time most of the Germans had ducked for the cover of their dugouts, others sprawled out in the bottom of the trenches dead or wounded, and a remnant still firing as we pulled away and headed for home. An exciting couple of hours with an excellent machine to fly in perfect weather, and a stout-hearted observer who was afraid of nothing. A man who said little but did much.

This particular show was successful. "Had's" observations compared favorably with the men who spent those long, solitary immovable hours on the ground. Theirs's was the tough job, but I was quite sure that what sustained them through those dreary hours of daylight, except those who went to sleep, was the thought of the man-sized issue of rum that would meet them on their return. Rum, during wartime, was part of the regular ration for those serving at the front in the British army.

Trench strafing, except for specified purposes, was up to the individual pilot. We were never sent out otherwise. But certain of us, especially those who had had a good dose of the trenches, welcomed the punishing advantage we held over these "friendly neighbors" across the way, and made the most of the opportunity. The risk was great but we were out to kill. An empty chair in the mess, and this came often, the scattered remains of a burning plane, and a charred body inside. But before we "bought it," we wanted to give it and the more the merrier.

Another form of decoy work which never fell to my lot was acting as a sitting duck for enemy fighter planes. With rare exceptions such as Ernst Udet, Oswald Boelcke and Werner Voss, the German fighter pilots seldom ventured across our lines even during their long periods of air supremacy. They believed in the philosophy of defense as against the offensive spirit of the R.F.C. and repeatedly refused combat. So, occasionally, we inveighed them into the open by offering them the most enticing of all baits – that ancient relic of pioneer days, the B.E. 2c, known by those who flew them as the bloody "Quirk".

The Bait – (B.E. 2c)

Soon after the Vimy show, Captain Al Ball, who was at the moment the leading ace, dropped into our Squadron No. 5 to visit his old friend, Gilbert, and to demonstrate the new S.E. 5-a which had replaced the Nieuport. His No. 56 Squadron was only a few miles away. They originally met at the Ruffy-Baurnann School in Hendon, the one from which I had also graduated. This established a bond between us and because of this, and Gilbert, he frequently landed at the field to gas up and load on more ammunition for another crack at the Huns, and chew the fat as he rested prior to takeoff. He hated the "Quirk" and all those responsible for its existence, having spent his first two months on the line flying one and knew that any German would go after one of these abortions. Even venture across the line, because they were so defenseless.

Inevitably, he and Gilbert got her heads together and on their own initiative decided to do a decoy job whenever weather conditions proved suitable. They said nothing but waited, and when the time arrived toward the end of April, they rendezvoused at a pre-arranged depot. It was not far behind the German lines and clearly visible to the troops on both sides so that friend and foe alike could enjoy the entertainment. Pilots, that is war pilots, often played to the gallery just as other sportsmen, actors, and politicians.

High above and up in the sun, Ball waited as old Gilbert floundered around in the skies like the fledging doing his first shot. As an added incentive he was doing exactly that and registering with effect. Eventually "Archie" signaled and seeing the burst Gilbert stopped the shoot and the gunners stood to and watched for the show as the flying antique flapped its wings.

Four Albatros D IIIs came tearing in, a new and deadly improvement over the D II models beginning to appear in force alone on the front. One could imagine their pilots laughing their heads off as the "Quirk's" nose went down and into the fatal dive in a hopeless attempt to shake them off. They followed, but before they could close in, down came Ball straight for the formation, splitting them up and in a matter of seconds almost, first one went down, then another, both in flames.

Too much for them, the other two broke off the fight and fled with Ball after them but they managed to out-distance him to live to fight another day. Ball and Gilbert returned home but Ball, who was just getting warmed up, sat around chatting for a while and then gassed and loaded up for another trip across the line. In less than a month he, too had gone the way of so many others but before he did, he sent forty-eight on the same journey. A boy barely turned twenty-one, he was second to none. Gilbert and his plane came out without a scratch, but he also bought it early in 1918.

Of all decoys, and I might say unintentional, for a brief period the Fokker Triplane was the most effective. Strongly resembling the Sopwith Triplane, it was repeatedly mistaken for such when it first arrived at the front by the pilots of the R.F.C. and its early successes were phenomenal.

Lieutenant Werner Voss first appeared with it over Ypres in August 1917 and within twenty-one days shot down twenty-two British planes, mostly R.E. 8's – an all-time record that was never exceeded. However, the temporary supremacy it gained through its close resemblance, rate of climb, and extreme maneuverability, was quickly lost and it was withdrawn from service by the end of the year due to structural weaknesses. In a dive the wings would collapse. Richthofen, however, refused to turn his in, so Fokker rebuilt and greatly strengthened his plane which he flew until he was shot down in April 1918.

I was almost caught in September 1917. This was on a day when the weather was perfect for photography and so I took off along with "Had" Driscoll to map some obliques [See Chapter 8] of several shoots our artillery completed a few days earlier. It was a ticklish job as we were well behind the enemy lines. Being able to take them at low altitude meant that the chances of being spotted by enemy fighters were considerably reduced.

Unlike those some pilots who shot across the line trusting to luck, we always cruised around on our side of the line searching the skies for signs of enemy planes before darting across. Only when it was clear did we cross. When we were well over the line, "Had," who was keeping his eyes glued to the skies, leaned over, shouted in my ear and pointed up above. There to my great relief at about 8,000 feet was a flight of "Tripes," six in all, flying in the same direction, but ahead of us.

This was certainly a stroke of luck, I thought, and felt completely secure. "Had," whose job it was to take the pictures, could now do so without worrying about enemy aircraft and come back with excellent shots. He wasted no time and with nothing to fear from enemy aircraft, he took special care in lining up his targets so there'd be no repeats.

We came under fire from "Archie" who was missing us by a wide margin as the shells burst high overhead and to our right. At the moment, I didn't give any thought to the erratic style of shooting from this particular battery which normally was pretty accurate and for whom we had a certain respect.

Anyhow, with an O.K. from "Had" I pulled the plane round, pushed the throttle wide open and with nose down and all the speed I could gather headed west for the field. Suddenly, as we drew close to our lines, "Had" grabbed me by the shoulder and shouting "Huns" pointed up. There directly above us and at about their original altitude were the Tripes. A second look and I realized what the excitement was about. Underneath the lower wings of each plane was the Iron Cross. There was no doubt they were looking for us, following the signals give them by "Archie" which I had put down to poor shooting. Fortunately, they never saw us and we made it across the line with safety.

Right: Hindenburg line. Courtesy of Wikimedia Commons.

Left: The D.H.9 was developed from the D.H.4. Its main improvement was moving the pilot back closer to the observer, which made it easier for them to communicate, an asset in combat. Unfortunately, the engine used was lower-powered and less reliable than those used in the D.H.4. Used as a day bomber, its modest performance and reliability resulted in heavy casualties. Courtesy of Wikimedia Commons.

Chapter 7: Observers-Corps Squadron

Observers received the same ground training as corps squadron pilots but with particular reference to artillery observation and direction. That was their responsibility in the early days of artillery cooperation. However, as enemy fighter squadrons increased in strength and the numbers of the loss of men and machines became too great, the pilot took over those duties and the observer assumed the status of an aerial gunner but cooperated on reconnaissance, aerial photography, and was in complete charge on contact control or any low flying patrols or missions. In the latter case, the Huns were the least of our worries as they considered it suicidal to attack at such low altitudes. And when the troops were going over the top, our ceiling was the barrage and no sane pilot would dare penetrate that.

On contact patrol the observer wrote the reports, marked off the maps, and decided when the message bags should be dropped. This was his show and the only one. The pilot had all he could do to keep the plane on an even keel and avoid shell fire overhead and machine gun fire from below. Other than that, the pilot was the captain of the ship and the observer his first mate.

Normally, the observer was an aerial gunner who watched the skies keeping a sharp look out for enemy aircraft and warn the pilot when they were closing in and then swing his guns into position ready to give them everything he had. It was plenty.

To keep them off my tail was his main objective should they be closing in. A quick change of direction or altitude could toss an observer out of the cockpit and did if he forgot to remain on the alert. Whether this young German lieutenant fell or was emptied out by his pilot is a matter of conjecture. In the German air force the observer was the man in command in a two-seater, the pilot being subordinate to him. Generally, a sergeant, the relationship between the two was often strained. In any case, with an open cockpit, there were more ways than one of tossing a man overboard and it did happen.

Above: Pilot dead beside his B.E.2c. Courtesy of the R. Duncan Collection at Southern Illinois University at Carbondale.

Above: Plane of Bishop and Allen going down in flames. Courtesy of the R. Duncan Collection at Southern Illinois University at Carbondale.

Tough as their job was there never was a real shortage of observers. In the early days of the R.F.C., unless a man was as fortunate as I was, it was customary that a man should first qualify as an observer before being permitted to graduate for pilot training and gain his wings. Incidentally, observers wore a half wing attached to the letter "O" and were appropriately termed "one-lungers." In the event that for some reason or other, they were "washed out" as pilots or not acceptable, they served as ground instructors, adjutants, or in some capacity on the ground. Many, once they had earned their wings,

Above: Nieuports starting. Courtesy of the R. Duncan Collection at Southern Illinois University at Carbondale.

Above: German Lieutenant observer found after falling from airplane. Courtesy of the R. Duncan Collection at Southern Illinois University at Carbondale.

Above: German pilot jumping from flaming plane. This is a photo of a model, not a real plane. Courtesy of the R. Duncan Collection at Southern Illinois University at Carbondale.

returned to France and gave a good account of themselves. Outstanding of all was Colonel William Avery "Billy" Bishop, a Canadian flying ace with a total of seventy-two victories.

The death of an observer and the circumstances under which he died for sheer agony and horror had no equal. He could be killed by machine gun fire in combat, or from the ground which was a merciful death; he could be smashed to a pulp when the pilot was killed or hopelessly wounded, lost control of the plane and crashed to the ground; or he could endure the most horrible of all deaths by being burnt alive less vividly reported in dispatches as "being brought down in flames" like this German. Frequently, they jumped over the side in the event of fire.

An observer could be killed and the plane and pilot return to its base. But with the pilot gone, his situation was hopeless. The parachute that could have saved him was available but not provided; and the dual controls which would have enabled him to bring the plane down in reasonable safety were not installed, but could have been. He didn't have a break but I never saw one flinch against the toughest odds.

Many lost their nerves and were "boarded," meaning that they were declared unfit for flying by a medical board and we never saw them again. Others returned to their regiments or transferred to other branches of the Army. The saddest of all were those whose nerves and minds were shattered beyond repair. Of these, their numbers were all too great. They were pitiful to see as they stumbled and shook in an everlasting nightmare and staring vacantly but without knowledge.

In all, the millions of words were written about the men who flew in the First World War but I have yet, in much of which I have read, seen any reference to one of the bravest of all those who fought – the Observer. He had more strikes against him than any other soldier no matter in what capacity he fought on land, sea, or in the air. I, and many other pilots owe our lives to them as they fought like hell to hold off the Huns. They kept their eyes ranging far and wide on the lookout, spotting them before they saw us and so giving us a chance to get the hell out of sight and with time to dive and hug the ground.

With the tremendous casualties suffered by the corps squadron pilots, too often they had to fly with the man who had just received his wings, unskilled and untrained in the elements of combat flying and the favorite target of the Hun. To spot a fledgling pilot was simplicity itself as they floundered through the skies in a series of flat turns. "Cold meat," as the German pilots called them, they were an easy mark. An excellent example was the death of young Allen described later in this chapter.

As I've already stated, many lost their nerves and among them were two who flew many hours with me. Men who I was sure had nerves of steel but under the strain they broke. They were real fighters. One other chap was an arrogant snob detested by everyone. He received a deep flesh wound from shrapnel as "Archie" caught us in a barrage of shell fire. We were glad to get rid of him, although he had plenty of guts. He had to sleep on his tummy for quite a while.

But even with the best of pilots, the observer was on the spot. For purposes of identification each pilot was allotted a number which was used when checking out the field and contacting a battery prior to a shoot. Eventually these numbers had to be changed when information came through that the Germans, through interception of our radio signals, had in the course of time secured the numbers of the more efficient artillery pilots and marked them for destruction which naturally included their observer. My number was seven until it became too hot.

By the time we became aware of this, in the early spring of 1917, the new series Albatros D III with two Spandau[1] front guns came into service and made the work of the artillery pilot more difficult and dangerous to such an alarming degree that crossing the line was suicidal more often than not. Although the machine slowed up the work we had set out to do, it was accomplished despite our heavy losses. But the strain was visible in the decreased morale as nervous breakdowns, now termed battle fatigue, became more and more prevalent. This was further strengthened by the failure of the government to provide the long-promised equipment. This was the end for many of our best men who were literally killed by the callous indifference of those in high places.

Through all this dark period, I had an observer, a young lieutenant from the Royal Horse Artillery, and Irishman from County Kilkenny or Cork. Anyhow, he was dubbed "K.K." or "Irish." Utterly fearless and a wizard with a machine gun, the tougher the odds the better he liked it, yelled his head off in a scrap but never lost it. To avoid jamming as much as possible he left a few rounds out of each drum but balanced this off by carrying extra drums. It appeared to answer the purpose.

I took him on his first artillery shoot where he put up a splendid fight and demonstrated his ability as a machine gunner against a flight of six Huns who quickly discovered they had a wildcat by the tail. At this time, they weren't aware we had switched from the B.E. 2c to the R.E. 8 where the observer was in the rear and equipped with two Lewis machine guns.[2] However, they broke off the fight as we neared our line as they often did when short on experience, and disappeared into the blue so the fight ended in a draw as a Neuport[3] came to the rescue.

With control of the skies being now in the hands of the Hun, a crack machine gunner was an invaluable asset and I decided to rely on "Irish." He was also a natural in the cockpit and no matter the angle or altitude of the plane he was impossible to shake loose, which was remarkable considering that most gunners have a hell of a time keeping their balance and handling the guns, or even staying there. He never wasted a shot but held his fire and always aimed for the pilot and the gas tank. He had every move figured out and eventually became a first-class fighter pilot.

The silhouette of the R.E. 8, except at close quarters, strongly resembled the B.E. 2c and until they became aware of the change, enemy fighter planes came in at an angle to the tail plane in the tried and proven approach. But now, with an expert machine gunner in the rear, the artillery pilot carried a sting in his tail. The "cold meat" had come to life and for a short period more than held its own. Unfortunately, most of the German losses occurred over their own territory out of visual range of our own troops on the ground and so no credit could be given for the one "Irish" brought down.

A change of tactics by the Huns, however, quickly offset the temporary advantage of surprise. Now the attack came from the front at an angle with the pilot as the objective and with, whenever possible, the sun at the enemy's back (the blind side). This was Baron von Richthofen's favorite approach which he impressed upon his own pilots. If the speed of the plane had been increased proportionately with firepower, we could have more than held our own. Unfortunately, the increase from 60 mph to 80 mph, with little improvement in maneuverability against the snap and precision of the speedier Albatros D III – 40 mph faster – left us in a precarious position as in the days of the B.E. 2c. In this instance, the observer was tossed out. Whether before or after he hit the ground is a matter of debate. But if he did then the death of the pilot could be attributed to the plane's falling out of control as he shows no indication of being shot. Having seen many die due to this, I am inclined to believe it was the cause of this pilot's death – further substantiated by the head wounds so plainly visible.

As pointed out earlier in the book, without a passenger, the R.E. 8 had to be flown using ballast – a sack of sand weighing 112 pounds being the usual replacement. What the potential result would be without this in the back was a thought that continually haunted me. But in my wildest dreams I never imagined a person would go over the side through uncontrollable fear and of his own volition just as a student will freeze on the controls as rigidly as the hand of a dead man.

In the summer of 1917, liaison between the Army Corps and he Corps Squadron was further consolidated by attaching artillery officers for a tour of duty with the squadron. These were men of battle experience including a number from batteries we'd directed from time to time. Before they completed their sessions, we took them up and gave them a practical demonstration of how a shoot was conducted. This was a specially arranged affair against a target within short enough range to permit us to operate from behind our own lines with the minimum of risk.

One man I took up was a captain attached to a battery of heavies with which I had done a number of shoots. He had been at the front for almost a year and looked to me like a man who could take it. Before leaving the ground, I discussed the shoot with him and also his attitude toward flying. Even the bravest men have been known to flinch at the very idea of leaving the ground. This was a normal reaction in those days from men who had watched too many shot down. He assured me he was anticipating the opportunity to witness the shoot, and knowing how to handle Lewis guns, he felt he could give a good account of himself if attacked. Satisfied he had what it takes, we took off.

As the shoot progressed, I would look behind from time to time to note how he was taking it and I occasionally throttled back to have a word with him, to keep up his morale. I even had him fire a burst or two into the German lines to make him feel he was also doing his share. By the time we'd finished the shoot I was quite pleased. I let him know we were finished and that now we were on our way down low so that we could get a close up of the trenches and the mud and filth of a battle ground that had changed hands several times in the course of the war. That was my mistake.

No sooner had I stuck the nose down to about 3,000 feet then he began to scream and shout like a maniac drowning out the roar of the motor as he yelled, "Let me out, Let me out!!!" I pulled back the throttle, brought the plane up and at the same time turned to see what the hell was going on. There he was halfway over the side and ready to dive. Horrified, or I might say almost petrified, I rose out of my seat and turned sufficiently to seize him by the belt, pulled him back into the cockpit and onto his stool. Then grabbing hold of my Very's pistol I belted him a couple of times on the head, knocking him out. From start to finish it was all over in a matter of seconds. Then reaching for the joystick I straightened out the plane which was flapping around like a wounded duck and made a beeline for the field.

After we landed he told me something snapped and he completely lost his head. All he could think about was to get out of the plane and away from the horrible nightmare below rising up to meet him. This is the amazing reaction he had. He actually believed the ground was only a few feet away and that he should get free before the plane hit. He was calmed down by the time we landed and other than a good sized bump on the head, was none the worse for his experience. He was pretty upset, however, and afraid the story might get around. But I told him not to worry on that score as it was an unwritten law of the corps not to discuss such occurrences.

Unexpected as the artillery man's reaction had been, "Irish" pulled the worst boner of all which almost cost me my life and for which I should have turned him in but for the strong liking I had for him and my admiration for his guts. So I forget it.

The old saying "All's quiet on the Western Front" is a truthful description of what did prevail in certain sections as the wholesale slaughter on the ground and the dog-fights in the air eased up. On one such bright peaceful and sunny day in the fall of 1917, I went up to do a counter-battery shoot and took "Irish" with me. Shortly before taking off, a plane came in from patrol and reported "No Huns and very little Archie." There were days when this particular "Archie" would only fire a few bursts and then take it easy. It was a good battery but within the hour I had them ranged and blasted hell out of them. So, we settled down to a nice comfortable patrol for the next two hours checking occasionally to see if our battery was still on target and dreaming about my next leave and the time I'd have in London, yet keeping an eye on the German back areas hoping to spot another battery in operation or other forms of activity. Occasionally, I'd glance back at "Irish" to see how he was doing and have a few words with him but as he kept his head down, I figured he was writing up his report so I decided not to disturb him, although he should have been watching the skies for Huns. That was my mistake and almost my end. Suddenly I heard the sharp crack of bullets tearing through wings and fuselage, smashing into my instrument board and one which took a slice out of the ear muffler on my helmet, as I found out later. With a quick glance over my shoulder and simultaneously yanking my ship into a climbing turn, I saw the "Irish" going into action as one Hun went shooting by. Three others coming in from as many directions were met by a blistering fire from "Irish" as I twisted and turned trying to keep them off my tail or getting me from underneath.

It was a peculiar situation as the Huns didn't dive on my tail one after the other as was often the case and them coming up underneath. Instead, they hung back, still pouring on the shots, and waiting for the other fellow to close in first. Then I knew they were mere fledglings. Anyhow, this gave "Irish" a break and he got a burst into one which took effect because he broke away from the other three in a dive and headed back east. Whether he made it to the field we never did know. Anyhow, the others broke off the fight for which I gave the Irishman full credit. He really burnt their tails and sent them on their way.

Arriving back at the hangar we wrote out our report, and after this was completed, "Irish" pushed off to his quarters and I walked over to the plane to inspect the damage. For a moment I thought I was going nuts for there on the bench alongside the fuselage waiting my inspection was *The Daily Mail* and a two-pound box of Fry's chocolates. The mechanics came across them as they were checking the fuselage for bullet holes, tucked away down the length of the fuselage behind the observer's stool. It seems that "Irish" had hidden them out of sight figuring what I didn't know I wouldn't worry about. He had overlooked the fact that after a scrap every inch of the plane was checked for bullet holes and other damage and then rechecked by me.

I told the mechanics to keep the loot, say nothing but keep an eye on the ship and that good old Kilkenny Irishman. Several hours later when "Irish" knew I was in the mess having a drink, he strolled over to the hanger, climbed into the plane to retrieve his precious possessions and began groping inside the cockpit only to find the cupboard was bare. He then looked around the hanger, in my office, and finally walked away no doubt cursing those stinking mechanics who were laughing their heads off watching like a pair of bloody thieves.

The incident was closed the moment I remembered the Hun going down and I never brought it up. That was a bad mistake. He was not cured and this time a promising young pilot by the name of Darcy was seriously injured. On what turned out to be his last flight, I sent "Irish" along with him on a day when weather conditions were perfect for Huns. A sunny day with drifting banks of cumulus clouds and wide patches of blue. Using the clouds as cover enemy aircraft would come darting out from behind them and any unfortunate R.E. 8 they connected with was doomed unless the observer was as expert as "Irish." Our casualties had been heavy within the past few weeks and I was trying to hang on to every possible pilot and especially those who were proving their worth.

Darcy and "Irish" took off, but before they did I warned my erstwhile friend to keep a sharp eye on the skies as the last plane in had reported plenty of Albatros floating around but had managed to duck them and avoid a scrap. Again, I slipped up, as did my mechanics, by not watching "Irish" close enough as he climbed into the cockpit. Within the hour word was flashed back from the front that a plane was down a few hundred yards back of the line, intact and apparently in good shape. The infantry had watched the fight – three Albatros – and reported our fellows putting up a good fight. They had driven them off but had to land as the pilot was wounded. He was already in the field hospital by the time I arrived. This was a pretty badly battered piece of territory, but by marvelous piloting Darcy had set the machine down on the only strip of level unbroken ground visible and in a hollow.

We drove up in a truck, myself and two mechanics, ready to dismantle the plane or fly it off. When we arrived, "Irish" was standing by looking very dejected, but not deflated, and surrounded by a crowd of infantrymen who were having a field day. Before we left the squadron, however, I checked out with Major Tyson and among other things told him that in all probability when I returned I'd have a copy of *The Daily Mail* and a box of chocolates to show him. He laughed, said I was crazy and to beat it.

After a quick inspection, to all appearances the plane looked in good shape and flyable so I told the mechanics to go over it thoroughly but first to remove the guns and ammunition and check and clean out both cockpits. "Irish" was standing on one side watching but said nothing. They suspected what I had in mind and so started on the rear one first coming up once again with a box of chocolates and *The Daily Mail.*

By now we had a good audience of boys from the reserve trenches nearby, so turning to a corporal I gave him the chocolates to pass around and also *The Daily Mail.* Why "Irish" hadn't done so before we arrived was hard to understand until I remembered he was a typical tight-fisted Irishman. Although nothing was said, he knew I had caught him after our own little escapade together but apparently didn't seem to realize how serious such behavior was, or he didn't give a damn. War has a peculiar effect on men and at times they can hardly be held accountable for their actions. But to deliberately endanger other men's lives by sheer neglect and indifference twice in succession was unforgiveable, especially after what was found in the pilot's cockpit.

Before going through the cockpit thoroughly, the mechanic gave a cursory glance around and then sat down in the pilot's seat to feel out the controls, moving the stick in all directions and pushing to right and left on the rudder bar. When he did this, he found that the stick jammed on right rudder. Reaching down to find the cause, he felt something wet and sticky, but solid. Overlooked by the men who lifted Darcy out of the cockpit – he was unconscious through loss of blood – was the pilot's foot inside his shoe but cut clean off above the ankle by machine gun fire that came in low through the side of the fuselage and passed out underneath. The mechanic was ready to puke as he held it up to show me. It gave me quite a jolt and "Irish" also, who I think was beginning to realize what a heel he was. I said nothing to him – he was all washed up anyhow – and we continued on with the inspection. There were numerous bullet holes that could be patched up later, but there was no serious damage to the plane and it was in flyable condition. Blood was all over the cockpit so before I got in I had the mechanics rinse it out thoroughly. There was no shortage of water as the countless shell holes were full.

However, the strip on which Darcy had put the plane down was far too short so, seeing there was no lack of manpower, I decided to have the plane hauled to a road running east and west about 100 yards away. This ran along the foot of a fairly steep slope at the top of which was our front line so we were free from observation by the Germans. Dumping all but sufficient gas to carry me back to the field and everything else removed in order to lighten the plane, a swarm of men got under and around and literally carried it across the shell-ridden ground to the road.

The road was pretty rough and battered but I figured that with "a flying start" I could lift clear of the ground before hitting a rough spot. Other than that, the road was free of shrubs and trees on either side and there was a stiff breeze from the west. But first we ran up the motor with chocks in front of the wheels – brakes were unheard of – to check. That was satisfactory with power to spare. So now came the "flying start." This is how it was done.

The chocks were left in front of the wheels with a long cord attached so they could be pulled away. Then as the prop turned over slowly, as many men as possible grabbed hold of the four rear struts, two on each side of the fuselage which braced the upper and lower wings. Then, with everybody holding on tight, I advanced the throttle until it was wide open and with the engine giving every last rev, the tail leaving the ground and the plane pushing hard against the chocks and ready to go, I then gave the signal and the mechanics pulled the chocks away. Now the plane was pulling at the leash, dragging the men slowly forward. Again at the signal, the men let go and with the tail well up and the air pouring over the wings, the plane shot forward as from a catapult and in a manner of feet the wheels left the ground, and I was up

Above: Aviatik C.II two-seater. Unusually, the observer/gunner sat in the front seat; flexible machine guns were mounted on the rails alongside the front cockpit. The pilot sat in back with a small headrest. The small rudder was not paired with a fixed fin and the flight characteristics suffered, limiting production to a single batch of 75 airplanes. Courtesy of Aeronaut.

Right: This Belgian R.E.8 was re-engined with a 180 hp Hispano-Suiza V-8. The additional weight of the engine, radiator, and coolant absorbed the additional 30 hp increase over the usual engine without any significant performance boost. Courtesy of Aeronaut.

Above: The Albatros C.III two-seater was widely used. Early models carried a flexible gun for the observer; later models added a fixed gun for the pilot for greater effectiveness in combat. Courtesy of Aeronaut.

and on my way. This was not as simple as it sounds and for many pilots it would spell disaster. A man must be able to feel the power of the engine, the tug of the plane as it tries to break loose, and time exactly the moment to ease it off the ground.

I was none too happy as I drew closer to the field to face a situation I hated to think about – reporting "Irish." But it had to be done. Good as he was and as much as I liked him, he was a dangerous man to have running loose. I sent him back with the truck. Luckily he was about due to go home, and after a lot of scrounging round between the wing and the squadron, Major Tyson arranged for him to be posted for pilot training without further delay. So, we ducked a court-martial and the R.F.C. kept a man who turned out to be pretty hot as a fighter pilot. We celebrated before he left, at his expense.

Darcy, the pilot, recovered, was fitted with an artificial foot, but never flew again. He had had a bellyful of R.E. 8's which cured him of any desire to fly. The last I heard of him he was on the staff at the "House of Bolo," the R.F.C.'s headquarters as we so aptly termed it. Respect for staff and ground officers was never a marked feature of the R.F.C. pilot except for those who earned their wings.

The punishment an observer had to take was rough. He had two major strikes against him before he ever crossed the line, namely, the pilot and the R.E. 8. As an experienced and proven observer, he would fly with the better pilots and generally was permitted to choose. But he also was sent along with the fledglings when first they came out to check and correct them in their work until they'd acquired the necessary skill and efficiency. Given a good pilot, the breaks were even; with a mediocre one he had a fair chance of survival. But with a man who, despite specially fitted glasses, was still half-blind, his future was hopelessly dim.

Such a pilot arrived at the squadron one summer's day in 1917 when the "flying circuses"[4] were hitting their stride and knocking off the R.E. 8s like so many ninepins. He did not last long. A week or two – so short a time, in fact, I never had a chance to talk with him. Before he went, he wrecked numerous undercarriages by landing several feet higher than where the ground was and putting the fear of God into his poor bloody observers. The C.O. was in a hell of a state, but little could be done about it as he was a member of an exceedingly influential family and more than a reasonable excuse had to be arrived at before he received the boot.

The day "Billy" Bishop bought it we were working the line together. I was doing a shoot about a mile to the north and he was on N.F artillery patrol spotting. This was an easy job because he didn't have to keep his eyes glued on a definite target and so was able to keep a good lookout for Huns. This was another one of those days ideal for enemy fighters, of drifting clouds when one had to be on guard every moment. His observer was a young lad of nineteen, Allen, who, having been with the squadron for several months, was looking forward to being posted to England any day for flight training. He was one of my close friends but unfortunately was in another flight.

About halfway through the shoot, three Albatros came diving in, made a couple of passes but met with such a hot reception from Jock Henderson, they shoved off which led me to believe they were merely fledglings. They climbed high above us and then turned south and went into a dive straight for Bishop's plane coming in from the west with the sun at their back. I don't believe Bishop or Allen ever saw them because by the time the second guy came in with both guns going, the plane was on its way down in flames from about 4,000 feet shot over as quickly as possible to watch its descent and spot where it hit. To my horror, I saw a figure crawl out on fire, slide down the fuselage a short distance, then fall off. It was Allen, a twisting, tumbling ball of fire that mercifully went out in a quick short burst of flame. Soaked with gasoline much of which they had swallowed, they burned like a torch in the breeze. What was left of the plane and two men who fell behind the German lines was never known.

Endnotes

1 The Spandau LMG 08 machine gun was an air-cooled version of the German Army's water-cooled MG 08 machine gun. It was based on the design of Hiram Maxim's machine gun. After its introduction in 1915, it was later made to function with the synchronizer system on the Fokker E.1 and led to a lengthy period of German air superiority on the Western Front. Baron von Richthofen used a twin synchronized setup. Weighing twenty-seven pounds, it fired a 7.92 caliber at 400/500 rounds per minute.

2 The Lewis machine gun was a gas-operated light machine gun built initially for the Belgian Army but was quickly obtained by the British War Office and manufactured at the Birmingham Small Arms Company's factor in Warwickshire. It performed well as an aircraft armament with a 97-round ammunition pan. It was used to shoot forward by the pilot or on a flexible mounting and used in the rear by the observer.

3 The Nieuport was a single-engine, single-seat French biplane fighter built by Société Anonyme des Establissements Nieuport which formed in 1909. It entered military service in the summer of 1915 and favored by many Canadian aces including Billy Bishop and Arthur Roy Brown.

4 In 1917, the Germany Army Air Service established the *Jagdgeschwader* fighter group. It was nicknamed the "Flying Circus" because the pilots painted their aircraft with bright colors.

Above: Albatros D.III fighter of *Jasta* 19 with its pilot, *Lt.* Reinhold Oertelt. He was killed in action by famous French ace Charles Guynemer on July 7, 1917. Courtesy of Aeronaut.

Above: Albatros D.II fighter that was delivered to *Jasta* 2. The D.III was developed from the D.II by adaptation of a "Nieuport-type" wing cellule. The higher aspect-ratio of the new wings reduced drag, gving the D.III better speed and climb rate than the D.II at the cost of a fragile, single-spar lower wing that sometimes failed during the stress of combat maneuvers. Courtesy of Aeronaut.

Right: Fokker Dr.I triplane. Its triplane configuration was inspired by the success of the Sopwith triplane. However, there was no commonality between the designs other than their triplane configuration and use of a rotary engine. The Fokker triplane was very maneuverable and had a good rate of climb, but it was slow due to its low power and high drag. The later Fokker D.VII biplane was a much better fighter. Courtesy of Aeronaut.

Right: The Sopwith triplane was the only British fighter of early 1917 that had better performance than the German Albatros D.III. German authorities thought the secret of its good performance was its triplane configuration; this started a 'triplane craze' in Germany that lasted until the end of 1917 and resulted in the operational Fokker Triplane and many triplane fighter prototypes that did not see combat. The Sopwith Triplane was replaced by the Camel. Courtesy of Aeronaut.

Right: Aerial reconnaissance camera mounted vertically on the left of a British F.E.2d with Rolls-Royce V-12 engine. The powerful engine anabled the F.E.2d to carry extensive armament and equipment; here it has two flexible Lewis guns for the observer and a fixed Lewis gun for the pilot in addition to its camera. The F.E.2b, powered by a 160 hp six-cylinder engine, was used as a night bomber to the Armistice after it became too vulnerable for day combat. Courtesy of Aeronaut.

Chapter 8: Aerial Photography

One of the most remarkable achievements of the allied and enemy air forces in World War I was the development of aerial photography. Within less than two years the camera became indispensable as a means of securing information of what was happening behind the enemy lines and supporting the reports of the reconnaissance pilots and observers by providing indisputable evidence of the results of the damage inflicted upon enemy batteries and other targets upon which they had been ranged. For mapping purposes, it was foolproof, providing as it did a continuity of photographs which, when pieced together, provided a scale map as accurate but far more descriptive of what was on the ground below than the most skillful and accurate map maker and surveyor could provide. This was accomplished through the medium of a vertical camera.

Periodically the front lines and the battle area behind were photographed by the Corps Squadron. From the maps built up from those photos and the resultant information, they determined much of the strategy and the tactics for an approaching battle. Nothing was hidden from the eye of the camera, no matter how high the altitude from which it was operated. Thick haze was no obstacle as filters to meet any condition were available. Pilot errors such as tilting the plane, were easily corrected and pilots were sent up again and again, if necessary, until successful.

Aerial photography was one of the most dangerous of all missions that a plane could be sent. Weather and Huns permitting, the favored altitude with the B.E. 2c was 7/8,000 feet and the R.E. 8 was 10,000 feet. At these altitudes "Archie" was none too accurate and the bulls eye was rarely scored. A few jagged shrapnel holes in the fabric would be the worst that could be expected. When the run was completed and the signal to break was given, it was every man for himself as he pushed his throttle forward and headed all out for home.

We had no rolls of film but used plates. The camera on the B.E. 2c was secured to the outside of the fuselage on the right-hand side immediately below the pilot's cockpit. The pilot snapped the pictures and changed the slides by hand as the observer sat tight on the alert for Huns. The full box of plates fitted on top of the camera and an empty box ready for the exposed ones was attached to the front and below the other. As each plate was exposed it was pushed forward into the empty box and the next plate to be exposed fell into position. Between exposures a time element of 8/10 seconds was counted and snap would go the plunger. The time element was calculated prior to takeoff by taking into account the strength of the wind at the acquired altitude and the air speed to be flown at. A rough and ready system, but it worked. This would permit an overlay of approximately ¼ to ½ inch on each plate. The space between planes also allowed for a lateral overlay. On landing the photographic section would take the boxes and within an hour the finished prints would be on their way to the mapping section at headquarters. There a mosaic would be built and distributed to the many and various commanders of artillery and infantry on the field and others of all ranks who required this information. For the infantry it showed a true and authentic outline of the enemy trench system they would eventually attempt to capture. For the artillery, an up-to-date scale map of the back area indicated possible battery positions, supply and ammunition dumps. And for the intelligence officer, it offered a special trench map to be filled in with the names of the German regiments secured as a result of trench raids or through spies operating behind the German lines.

Only on rare occasions did a photography flight return with a perfect set of pictures or the men they started out with. The path to success was strewn with too many obstacles – weather, enemy aircraft, the everlasting engine trouble of the R.A.F. designed engines, and finally, bad piloting induced by fear as "Archie" filled the skies with a barrage of black smoke and shrapnel. A pilot who tried to dodge and break formation was bad for us but worse for him. The gaps he made had to be filled in and his was the honor to whom it was given. Worse still, he had to do it alone. Sometimes the job was completed; more often he never returned. Our losses were fantastic and most were unnecessary as the Albatros and the Fokker Triplanes reaped the harvest moving across the skies in the "circus" formations of the Jagdstaffen.

Outnumbered as we were, even with our fighter escort which we rarely enjoyed in the bad days of 1916/1917, it was a dead certainty that our photography was over for the day as they came in and we fought for our lives. It was every man for himself as the dogfights quickly took shape into individual scraps with planes twisting and turning in every possible attitude in a desperate effort to get their man or defend themselves. The scene is an excellent illustration of an Albatros with one wing folding back and on fire, passing below the R.E. 8 but caught in time by an S.E. 5. Whenever possible, our fighters never failed to come to our rescue.

The R.E. 8, despite its manifold faults, was a good photography plane, inasmuch as the camera was now installed in the rear cockpit and operated, but not completely, by the observer. The pilot still controlled the timing, indicating the moment each plate should be exposed. This presented the observer with the difficult problem of operating the camera, keeping one eye on the pilot, and being on the alert for Huns.

When carrying out this work, "Archie" was well aware of what we were doing and invariably started ringing us with a barrage of shrapnel and black smoke hoping to bring the plane or planes down, and at the same time, signaling

Above: Photograph of models illustrating a dogfight. Courtesy of the R. Duncan Collection at Southern Illinois University at Carbondale.

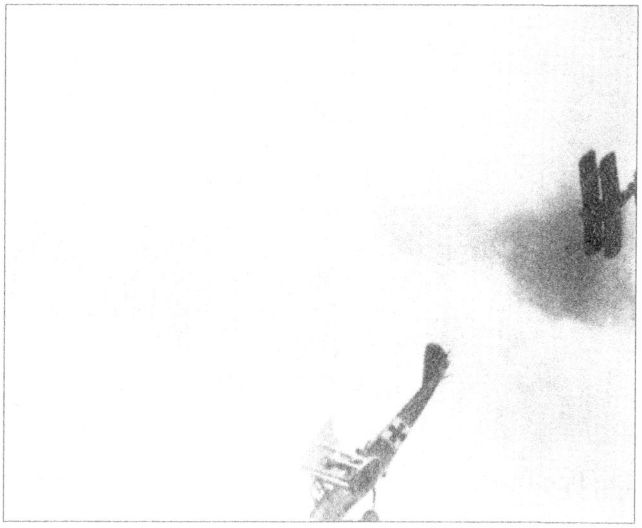

Above: British fighter attacking a German Rumpler reconnaissance two-seater. Courtesy of the R. Duncan Collection at Southern Illinois University at Carbondale.

Right: Very early production Rumpler C.IV reconnaissance two-seater. The straight observer's gun mount was soon replaced by a circular, rotating gun mount in production. Courtesy of Aeronaut.

Right: Rumpler C.IV 6692/16 reconnaissance two-seater. The high-flying Rumpler C.IV was Germany's premier long-range reconnaissance plane; its high speed and ceiling made it very difficult to intercept on these missions. Initially powered by a 260 hp Mercedes D.IVa, later variants were powered by high-alititude Maybach engines for even greater performance. Courtesy of Aeronaut.

This Page: Light and heavy German anti-aircraft (FLAK) guns. Courtesy Aeronaut.

to enemy aircraft in the vicinity. We paid little attention to "Archie" but we were alerted to the danger of a sudden attack. If caught well over German territory our chances of survival were limited unless our own fighters came to the rescue or we were able to put up a running fight long enough to stall them off until we reached our own side of the line, clouds came to the rescue, or the German obsession against carrying the fight across our line.

They did on a day in the late fall of 1917 when intermittent patches of clouds began to drift across the area between us and the ground as we were photographing the results of several artillery shoots recently carried out by several long range batteries on a number of strong points and battery positions.

We lined up our objectives and the flight was going smoothly but delayed somewhat by drifting clouds blotting out our objectives. Typical Hun weather, they made good use of clouds, but so far there had been no signs of them. "Archie" kept tossing up a flurry of shells now and then but other than that the flight was going along smoothly.

With less than an hour to go and the job almost completed the chances of tangling with the Huns were becoming more remote by the minute and I was getting ready to ease down to a much lower altitude for a few oblique shots when I felt that old tap on the shoulder.

Automatically, I yanked the machine round into a vertical bank offering the least possible target and then looked around at Henderson who pointed down and at an angle. As I glanced over the side and in the direction he indicated I received as jolt – a flight of seven Triplanes were only a few hundred yards well below us but rising fast at a terrific speed. On the top wings was painted the Iron Cross. For a moment, I assumed they were from No. 10 Naval Squadron who had come to my rescue more than once, but a second look told the story. There they were, as they came into the sun, part of the good old "circus" with all its colors splashed over the planes coming in. I was surely in a tight spot as I knew once they closed in and began shooting I was a dead pigeon. Actually a dogfight is a whirling mass of planes going in every direction, making it impossible to distinguish one's opponents. I had a good look at the planes but all I could see was a mixture of vivid colors. But, of course, I didn't waste much time looking.

Anticipating something like this, I kept an eye on the clouds, and a couple of hundred yards away were two floating slowly along and pretty good sized ones. Without a moment's hesitation, I pushed the throttle forward, stuck my nose down into almost a vertical dive and went into that cloud expecting at any minute to hear the wings ripping lose from the fuselage. Once inside that mass of whiteness, I eased back on the stick and after what seemed an endless journey, came out at about 2,000 feet above the ground. They didn't follow me through and I ducked that potentially fatal scrap without a shot being fired by anyone. I looked back over my shoulder at Jock and there he was, not the least bit disturbed and with both guns ready for action.

We had come out over our side of the line and after cruising around for a few minutes, went back to work on the obliques. In doing this we occasionally came under machine gun fire that caused us little worry. Later, when we arrived back at the squadron we learned we'd crossed paths with the Baron von Richthofen whose bright red plane the troops on the ground had been able to spot. For close up shots of ground objects such as barbed wire, trenches, pill boxes, etc., an oblique camera is employed. In appearance a small copy of the vertical camera, it is grasped in both hands by the observer for snapping shots at an angle from the side of his cockpit, hence its name. Designed for low altitude work, it is an excellent source of information in picture form. Also, for confirming the visual reports of individual pilots. However, it is confined to low altitude operations and considered to be, due to the heavy concentration of machine gun fire from the enemy trenches, one of the more hazardous duties of a corps squadron pilot.

The vertical (mapping) camera was put to use with amazing results. Flying at the lowest possible altitude at which the camera would remain in focus, stereo shots are taken from which the interior of a trench, battery or any ground objects would stand out in sharp relief when viewed through a stereoscope. The methods were simple but as the operation was invariably carried out under heavy machine gun fire, it demanded a total indifference. It required an absolutely level flight – not tilting in any direction – and the exposure of two plates with the utmost speed. The changing of plates was a manual job and the observer who did the shooting had to work fast. Practice developed speed. The hardest part of the job was focusing on the targets under heavy fire. I'm happy to say I never lost and observer as a result of it. And one reason, and about the only one, was the preliminary rundown and the strafing given them before we actually began snapping pictures. Nothing better confirmed General Trenchard's theory that the offensive is the best defense.

In northern France, photographic work was slow and arduous. September until early summer was a period of heavy continuous rain and fog. The ground below, with the ruination of its drainage systems, had returned to the original swamps of centuries ago and no doubt contributed to the abnormal conditions which prevailed throughout the war and hampered our work.

Delayed by weather, which would change within a few hours or within the hour, high altitude flying essential for mapping purposes was one of the more hazardous tasks of an artillery pilot. Blinding fogs would roll in when least expected and the area to be mapped was blotted out. As a consequence, many planes and their crews were killed as they crashed into the ground they were unable to see. Others lost their bearings, flew in the wrong direction, and seeing a clear opening ahead, landed only to find themselves in German hands. Photography was expensive in men and machines, but regardless of the losses, we kept at it until we got what we were after and the maps were completed.

Above: Heavy German anti-aircraft (FLAK) gun after firing. Courtesy Aeronaut.

Above: Heavy German anti-aircraft (FLAK) guns. Courtesy Aeronaut.

Chapter 9: Contact Patrol

In the early days of the war, adequate contact was maintained during the course of a battle using telephones and runners between regimental, brigade, and divisional headquarters. However, as the war progressed, and artillery on both sides increased in power and strength, and the barrage which was used to cover the advance of an attacking force was supplemented by another more destructive, of greater intensity, extending in range and depth to areas far beyond the scene of hand-to-hand combat. The artillery in this barrage were not the light field pieces used to cover the advance of the infantry, but comprised the heavies and long-range guns of all calibers. Theirs was the job of seeking out and endeavoring to destroy opposing batteries and strong points, and cutting off and wiping out all or any support such as reserves, supplies, ammunition – even first-aid stations – and so halt or slow the progress of the attacking forces. Under this tremendous hail of shrapnel and high explosives, including gas shells, the ground was torn asunder, telephone lines were obliterated, and the use of runners was suicidal as well as time consuming even should they get through.

To resolve the problem, the airplane was brought into use supplanting the old methods by providing a comparatively safe, speedy, and accurate way of maintaining contact with the most advanced troops and the various headquarters to the rear, however distant.

This was described as Contact Patrol. Through this method of communication, the progress of the battle along each foot of the front was under constant observation, even the most isolated spots. Not from a distance as in an observation balloon or through the field glasses of some observer searching the scene from a solitary post, but through the eyes of a plane directly overhead, moving forward with the troops with two skilled men watching and recording what they saw and leaving nothing to guesswork.

A battle is a scene of men running, some dog-trotting, and others plodding slowly forward, or backing up amidst the rising smoke befogging the ground from the shells tearing it and the moving men apart. Shuttling back and forth, a single plane could furnish a continuous series of reports to headquarters of what was occurring in every section of his particular battle area. They flew so low, often a hundred feet or less, meant they could distinguish the uniforms and equipment without confusing them with the enemy as the more distant observers frequently did.

Contact with the troops below was maintained by a system of signals throughout the duration of the battle. Messages as soon as they were received were on their way as fast as the plane could fly and rarely did they fail to get through unless hit by our own shell fire or enemy machine gunners. For purposes of identification, and to avoid the possibility of confusion with other aircraft of either side, every contact patrol plane carried special markings. These were provided by streamers attached to the inter-wing struts and black flaps fastened to the trailing edge and at the center of the lower wings.

Method of contact or communication in actual battle was by means of a Klaxon horn tapping out in code, or varied-colored flares. They, in return, contacted us through the use of what had the appearance of an oblong length of green canvas lying flat on the ground. This actually was a venetian type affair which, when the flaps were open, displayed a white background. Using Morse code, they wig-wagged the messages at a speed of about eight words a minute. By no means a record speed, but compared to the late lamented runners it was like a flash of lightning.

When the flap was not available, ground strips were resorted to, or signal flags. To read all these and decipher them put quite a train on the observer and it was nothing unusual for him to become confused. All being well, however, within a matter of minutes after the message was received, we were dropping it in a circle near whatever headquarters was assigned for this purpose.

There were no walkie-talkies in those days, but crude as it may appear, it worked. The message was tucked away in a small canvas bag, lead-weighted, and to which were attached long, brightly colored streamers.

At headquarters a circle was usually marked out on the nearest available open space and as free from obstacles as possible. Coming in low, we would drop our message, trying always to hit the bulls-eye, fly around until they had read it in case of reply, and then – given the okay – off we would go back to the big show. To put a little humor into it all, and when we were feeling our cats, we would, on occasion, fly straight at the guy waiting to pick up our message and laugh our heads off as he hit the mud. But it was all good clean fun, as they say, and it gave the General a good laugh and a moment's respite from the horrors of war as he sat swigging down his Scotch in the warmth and comfort of an old French farmhouse. It was also a comforting thought for us that we would be following his example before dusk set in as we headed back for more trouble.

Normally the messages called for support, supplies, or that ammunition was running short along with the supply of Mills bombs. These latter were tossed around like baseballs. Shaped like a lemon they were extremely effective. Other messages would concern medical aid, and on rare occasions, indicated the barrage was falling short as our men advanced too fast.

Contact patrol for an inexperienced pilot was considered one of the riskiest types of work a wartime pilot was called upon to perform. I do not speak of the recently arrived fledgling pilot but refer to men of reputation but unpracticed in the particular skill this type of flying called for.

Even an average pilot, given plenty of altitude, can put

a plane through its paces. Should he make a mistake, and on doing so lose altitude, he has plenty of room in which to recover. Tree-top flying, however, is a vastly different story at an altitude of 100 feet or less. A slight stall or slip may spell disaster unless there comes that subconscious warning and quickness of control, an automatic reaction that only occurs with continual practice and which is precision flying in its highest form. Having this, which is 90% of the business, the simple flight manoeuvers employed provided a background in which the chances of return were more than even.

The approach, once within range of the artillery duel, was made on either flank of the sector of operations and the altitude at which the plane was to be flown. Then cruising along observing and checking as the battle progressed, manipulating the rudder a fraction slightly to the left and right, a skidding action was set in motion like a car on the road, rendering it difficult for machine gunners or marksmen to take aim with any degree of accuracy. There was little danger of falling off into a spin as the balance of the plane was unaffected, but the continual change of attitude served its purpose.

There was nothing to be feared from anti-aircraft fire as their trajectory was too high but we did face considerable danger from both our artillery and the Germans as they dueled it out. With thousands of weapons hurling shells of all calibers, ranges, and trajectories into the air – a positive blizzard of steel – the risk of a plane being hit was always to be reckoned with. This determined the altitude at which the plane could be flown with comparatively safety. Normally, between 100/200 feet above the ground. At this height the plane was sandwiched between the flat trajectory of the creeping barrage behind which the infantry was advancing and the much higher one of the heavies and long-range guns. Theoretically, it was an ideal selection, but like anything else, there was the occasional slipup. Luckily for the pilot it was a quick death and not the slowly agonizing but all to frequent one of coming down in flames.

That a pilot and plane were within the danger zone was typically indicated by a series of quick short bumps from the disturbed air caused by passing shell – a warning that it was time to move elsewhere. Once, during a shoot with about the largest gun operating along the Western Front – a 12-inch Naval gun on a special railroad mounting – I saw a brown blur pass close by my wing tip and simultaneously received a terrific dump. I moved over in a hurry and stayed on my new course until the target was demolished.

German machine gunners were well trained, as the German High Command appreciated the value of the machine gun and the tremendous firepower it possessed. So every man was almost as accurate when firing at aircraft as on the ground, provided the machine gun moved along in steady flight, followed a given path, and within an altitude of 1,500 feet. This was demonstrated unfortunately in the death of Major Edward "Mick" Mannock V.C., the top British ace with 61 victories to his credit. He was duck soup for them despite the fact he was a marvelous fighter pilot. But he knew nothing of the dangers that lurked close to the ground. Curious, indeed, that the two greatest aces of the war should die in such an unfitting manner. Captain Albert Ball was the other.

On the day Manncock died, he took up a young pilot to teach him the trick of getting a Hun. They soon found what they were looking for and Mannock maneuvered him into position and the young pilot shot him down, but not until they were close to the German front line and at a fairly low altitude. The infantrymen watching the fight reported that as Mannock crossed the line they heard a burst of machine gun fire and down he came. He was flying as an S.E. 5, one of the latest model fighter planes with a speed of 130 mph and next to the Sopwith Camel the most efficient and deadliest of all allied fighter planes.

Working along an extended front such as Arras – a 14/20-mile stretch which included Vimy Ridge and involved the 1st and 3rd Armies – several contact patrol planes were operating at the same time and care had to be exercised to prevent overlapping or wandering into another man's territory. In the heat and excitement of battle anything can happen especially when it becomes a long drawn out affair as this did. It began on April 9th and both sides battered and hammered one another for the next two weeks.

Pilots and observers were furnished with trench maps of recent date, reproduced and drawn to scale from a mosaic built up from thousands of photographs of the German trench system and the area immediately behind. On these maps were marked off the various objectives our troops were driving for and other information of assistance to pilot and observer. They covered his section and field of operations only and so avoided the dangers of overlapping and conflicting reports. Equipped with maps, message bags, writing pads, pencils, Very pistols and flares, and Klaxon horns, we would take off and head for the line to cover the the men below.

Keyed up and on the alert, we concentrated on nothing but the fight, watching groups of men advancing and stopping, moving on again, stumbling and falling, and not getting up. Nothing appeared to make sense, a crazy mob of squads going nowhere to all intents, but they were on their way. And up above we followed it through, and with flares going up, we marked off this spot and that and finally it began to make sense as the lines joined up. The observer was a busy man.

My job was to fly the plane and follow the course of the battle. Close to the ground the air is rough and disturbed by the never-ending barrage of shells passing overhead and a pilot has to struggle hard working the controls, watching the instruments, and the ground below. This is not flying but sheer hard work where the slightest slip of hand or foot will send the plane hurtling to the ground and oblivion.

We are both on edge, but not scared as by now it's old stuff but the tension is always with us and always will be. But it's exciting and then comes relief as Jock leans over

Above: British troops adavncing with tanks at Vimy Ridge. Courtesy of the R. Duncan Collection at Southern Illinois University at Carbondale.

Above: Medical Dressing Station during battle. Courtesy of the R. Duncan Collection at Southern Illinois University at Carbondale.

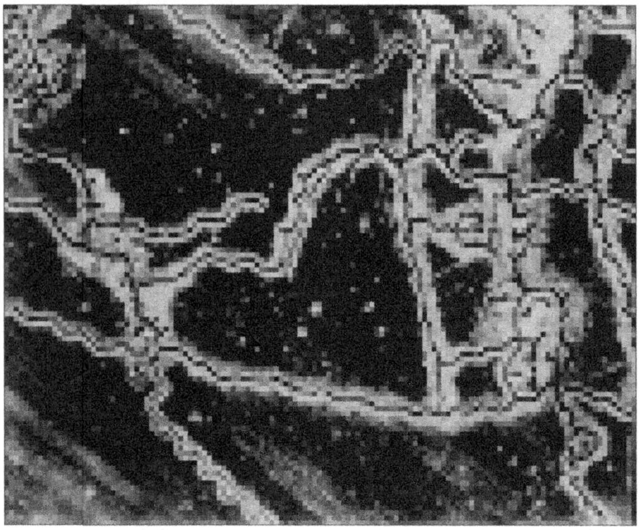

Above: Aerial photograph of trenches. Courtesy of Wikipedia Commons.

Above: Irish flying ace Major Edward "Mick" Mannock in the R.F.C. Courtesy of the R. Duncan Collection at Southern Illinois University at Carbondale.

Left: The Sopwith Dolphin arrived at the Front in early 1918, the next Sopwith fighter after the Camel. Powered by the same 200 hp Hispano-Suiza engine used by many S.E.5a fighters, it performed well at altitude. Despite its good performance, it was not popular with its pilots due to their exposed heads in case of over-turning.

and shouts in my ear – "Let's go with the bag." I yank the plane round on its wing tip, level off and make a beeline for headquarters. We arrive at the dropping circle, but there's nobody standing by. So we toot on the Klaxon horn and out comes a man with the red tabs – a staff major. He picks up the message, gives the "Hi" sign and away we go past a stream of wounded on their way to the rear. We give them a wave and a couple of toots on the Klaxon horn and then back into the battle.

At Vimy Ridge we had a real tough deal. We were the Corps Squadron to the Canadian Army Corps. They went over the top in the face of driving sleet and snow and we went with them. Visibility was poor and made our job more difficult but failed to obscure our view of the battle below. It was the morning of April 9, 1917. Slowly the men pushed forward and along with them a scattering of tanks, the first we had ever seen but too few to be really effective. Finally, the weather cleared and the whole picture slowly unfolded. Two armies were on the move along a fourteen-mile front with Vimy Ridge only a small but exceedingly important part of it. Ours was the northern flank with Arras further to the south. But the whole panorama was lit with thousands of shells bursting, tearing men and ground to pieces, burying them never to be found, obliterating all in their path and covering the land with drifting clouds of smoke, and columns of mud and water spouting to the skies in countless geysers.

To us it was like watching a silent movie. We saw it all but heard nothing over the roar of our engine. It was like a nightmare, to see and not to hear, as of one being there and yet not there. To shake it off and toss the feeling aside, I'd throttle back and a steady drumming sound like muffled drums could be heard. But nothing of the clash of bayonet against bayonet, the burst of hand grenades, the staccato rattle of machine guns and the screams of the wounded and dying. Visible also were the vicious flashes of thousands of machine guns and rifles as the enemy held onto pill-box, machine gun nests, and strong points. The Germans had two years to prepare for this show and what they held was a fortress in length and depth with the bomb proof shelter of dugouts three stories deep and covered with solid concrete.

Talk about the greatest show on earth, this was it. And together, my fighting Gordon Highlander, Jock, and I were watching the biggest and the bloodiest, 132,000 casualties, as we cruised up and down and never higher than 200 feet – more often at 100 – and a sight given too few to see from such a vantage. Cruising free as a bird, darting from one place to another as we followed on and over, noting this and noting that, thanking God that we were high and dry. Watching the bombs fly, the glint of the sun on bayonets as men fought hand to hand, the quick darting flash of the machine guns and men, dirt, and mud flying up in the air as a shell landed in their midst. We were there and we saw it all.

The artillery preparation had been exceptionally good, the wire in front of the enemy trenches had been swept away and scattered leaving wide open gaps through which the troops moved freely. The first line of trenches was flattened but back beyond them as the defense began to stiffen, the barrage lifted, rolled onwards, and the men slowly advanced behind its cover. It was all a marvelous demonstration of planning, timing and cooperation, dogged resolution, and guts by all ranks.

After the first hour we knew we had it won but the excitement of battle and the weakening defenses of the Germans caused one body of men to go ahead faster, much faster than they expected, and capture a strongly fortified village, a key point of major importance to the enemy. In doing so, they'd lost contact with the rest of the troops on their flank and suddenly found themselves hemmed in by an enemy determined to regain control.

The first indication we had of the situation were a series of flares, shooting up into the skies in quick succession. With tail trimmed and throttle wide we shot over in their direction and swooped down, circling the ruins of a village to find it packed with Canadians who in turn were completely surrounded by Germans. Meanwhile, we came under heavy fire with everything coming at us but the kitchen sink, including as the Canadians stated later, trench mortar shells.

Without signaling back and forth we quickly grasped the situation, checked their position on the map, flew over to brigade headquarters and dropped the message bag with map and report. We circled the spot awaiting the reply which was quick and to the point. Relief was on the way. Tell them to hold on. Back we went to the village, dropped the message which Jock had written out and went back to the line with nothing more than a few bullet holes in the wings and a thirst that was choking me. Incidentally, the officer responsible for capturing the village was an old friend of mine with whom I'd served in the C.M.R.'s. He was killed.

The Canadians finally occupied the Ridge which overlooked the plains surrounding the city of Lens. It was a severe blow to the Germans who fought hard to retain it, suffered heavy casualties, and had thousands of their men taken prisoner, including guns of all calibers, ammunition, and supplies.

Before turning for home we were able to get in a good lick in ourselves at the enemy, raising hell with men and horses as we shot them up as they retreated down the roads and trails leading to their new Hindenburg defenses. Risky work but exciting with the only fear being that of the engine conking. On these shows, Jock was the lad to have behind me. He'd scream and curse the demoralized troops as he sprayed them with his machine guns.

Well satisfied to be heading back to the field, we came down low and flew alongside the long lines of wounded making their way back to the dressing station waving and blasting away with our Klaxon horn. That, I knew, really gave them a lift as they waved back frantically. Then, as we neared a Model T ambulance, Jock shot a message over the side to the boys below addressed to the doctors, inviting them to our mess whenever convenient. They came and we gave them a royal reception and a riotous evening till dawn.

The following day several of us piled into the flight tender and drove up as close to the front as possible. We stopped at a ruined village about two miles behind the old front lines as there was nothing but shell holes and churned up ground ahead. Further travel was impossible except to pack mules which moved up supplies to the troops far in advance. A dump had been established in the village and a couple of long-range batteries had also moved in but they still lacked ammunition. The rest of the artillery were unable to move forward until roads were built. This would take several days and was one of the main reasons the advance could not be followed up with the heavier guns so necessary to keep the enemy on the move.

We left the tender and struck overland skirting the edges of shell holes and craters, stumbling along over piles of battle debris until we came to the edge of the ridge overlooking the plains stretching away in the distance and the new lines of the Canadians which were gradually taking form. Sitting down nearby was a Canadian soldier with his rifle resting against his right shoulder and staring out across the plains. I walked over to speak to him but he didn't answer. Then I looked again and saw he was dead. He was the picture of health with not a mark on him. We left him alone wondering what the manner of death had his been. Bodies were lying everywhere but being carried away as rapidly as possible. So far a party hadn't reached him and he was waiting.

After rambling around the German lines which were flattened, we left quite pleased with our artillery work, picked up the tender and drove off. Within the hour after we had gone the Germans opened fire on the village with long-range guns firing from well behind their new position – the Hindenburg line – and literally wiped out the two batteries, the dump, and killed a number of men.

Above: The brawny D.H.9a was developed from the earlier, under-powered D.H.9 by installation of the 375 hp Rolls-Royce Eagle or the 400 hp Liberty engine produced in the US. Both V-12 engines provided much greater power and reliability than the 6-cylinder engine in the D.H.9. The Liberty engine was almost the only American aviation success story of WWI. A heavier airplane than the D.H.9, it had a larger wing span and area than the D.H.9. Its greater power and larger wing gave it far better performance than the D.H.9 and it suffered much fewer casualties during its day bombing missions. Its strong performance made it the second best day bomber of the war; the French Breguet 14 B2 was the best day bomber of the war due to its metal airframe. The wood airframes of D.H. aircraft caused many casualties due to wood splinters during crashes. Courtesy of Aeronaut

Chapter 10: Cavalry Contact Patrol

The Battle of Arras of which Vimy Ridge was a part, continued and with it we were given a new slant on the activities of the contact patrol pilot. Thoroughly convinced this show would provide the long hoped-for breakthrough, a brigade of cavalry took up position not too far ahead of the scheduled zero hour. Attached to them was a flight of Armstrong-Whitworths (Ak-W's) as cavalry contact, something about which neither they nor the cavalry knew very much, if anything.

The day came and the hour struck. Over the top went the infantry and the 3rd Army was on the move to rival the glorious victory the Canadians of the 1st Army on their left had scored. General Sir Edmund Allenby, the G.O.C., a bad tempered and domineering cavalryman but a fine soldier, was the genius in command. The troops advanced a matter of a mile or so. Then, that night, without surveying the ground, he telegraphed orders from headquarters, miles away from the front, for the cavalry to be sent in the next day. The charge of the Light Brigade was about to be repeated.

They started off with a flourish but degenerated into a stumbling, halting movement as the horses wallowed and fell in the mud and slime and the thousands of shell holes blocking their path. From the shelter of pill-box and strong point, the krauts poured a murderous fire into their ranks mowing them down, killing horses and men until they lay in heaps along the battlefront. The advance was halted and the few that were left staggered back towards the rear.

Overhead was the contact patrol plane piloted by a South African by the name of Schwartz. Realizing the impossibility of the situation and hoping to stop more cavalry from being massacred, he swung his plane around to head for headquarters and drop his report when his engine was hit and the magneto shattered. Down he came, the plane was wrecked but neither he nor his observer were hurt. Climbing out of the wreck they found themselves surrounded by rider-less horses, half-crazy with fear and excitement. Schwartz, who was also a cavalryman, grabbed one, mounted and caught up with the remnants of a once proud cavalry regiment. Again disaster struck. He hadn't gone far when his horse was shot from under him.

By now he'd had a bellyful and so, trudging back to his plane, which he'd left a good mile behind, he picked up his observer and taking the two rear Lewis guns with them, they hiked another two miles to the ruined city of Arras before they were able to get transportation back to their squadron. He turned in his report to corps headquarters and that was the end of the cavalry show for him. He'd certainly pulled a boner and a stupid one. He didn't appear to understand that his job was that of furnishing headquarters with the progress of the battle and getting his report there with the utmost speed, particularly when he saw the predicament the cavalry was in. Instead of going ahead with the cavalry he should have headed the other way. His "heroic" gesture ricocheted with a vengeance.

Back at the squadron he wrote out a voluminous report of what had occurred, recommended himself for the Military Cross, and took it into the C.O. personally for him to read. Normally, the C.O. would shove it into the orderly room for the intelligence officer to handle and who would then return it to the C.O. if he thought it necessary. Then the C.O. would head for the mess and drink.

This being his first job over the line and his first time under fire, Schwartz frankly believed his exploits were the greatest and deserved even better recognition. The C.O. quickly cut him down to size. After reading through his escapades, he looked up at him and said in his quiet but compelling manner, "You know I don't really believe I should recommend you for the M.C. Frankly, you're damned lucky not to be under arrest for horse stealing." And, as they say, that was that.

Right: Typical wood and canvas two-seaters suffered many casualties from ground fire on their low-altitude infantry cooperatioin missions. Germany developed a special category of armored aircraft to survive these missions; the Junkers J.I was the best of these types and was the world's first all-metal production aircraft. Courtesy of Aeronaut

Chapter 11: Artillery Directions and Controls

The major task of the R.F.C. was artillery direction and control. Without adequate preparation and support from the artillery a successful attack or "push" was impossible. A tremendous responsibility, therefore, rested with the two men in the plane above, who, if through one cause or another, failed to do a thorough job of ranging the batteries they were directing, a "push" could be halted in its stride, thousands of men wcould be sent to their deaths needlessly, casualties would be high, and an army reduced to a fraction of its fighting strength. Mostly poor direction was caused by lack of experience which time alone would correct.

Gradually a system of checks was devised and inaccurate ranging was speedily corrected. Every ranging shot and its direction and distance from the target was written down on a special pad, and where doubt existed, a photograph of the shot would be taken. Some pilots were lacking in judging distance, others in a hurry to finish a shoot would give the battery the signal to go on firing in their own time even when the guns were way off target. Photography or a repeat shoot by an experienced pilot would decide whether negligence or lack of judgment in sighting were contributing factors. Sometimes the battery commander would be at fault, particularly should he and his men be comparatively short on battle experience.

Artillery direction and control was concerned mainly with the destruction of enemy batteries termed counter-battery work. The procedure followed in carrying this out was termed a "shoot." A good "shoot" was when we ranged the guns right on target and kept them there by accurate direction and a battery of first-glass gunners. A bad "shoot" may be caused by sloppy ranging from the air due generally to inexperience, or the battery may be at fault for one reason or another. Worn-out guns were a source of considerable trouble during the first two years of war.

Originally, artillery direction was under the control of the observer but as the enemy grew stronger in the air and the mass formation of the "circus" became standardized, the pilot took over and the observer was relegated to the role of machine gunner. His job was to keep his eyes glued to the skies on the alert for enemy aircraft, hunting for what they called "cold meat." Sunny days were the worst as the Huns would take advantage of the sun and attack from the blind side, and as artillery fire was from west to east, the plane would be flying in the sun a good part of the day. Incidentally, sunglasses like parachutes weren't around then.

Preparations for a shoot were simple. Once we were given the target we marked it off on gridded maps drawn to scale, noted any landmarks that may be of use in judging distance and then with maps, writing pad, and pencil, were ready to go. For localizing the shots, we employed the Clock Code – the numerals of the clock indicated the angle of the fall of a shell and certain letters the distance. Z was zero, the bull's eye. The center of the clock and many bull's eyes we scored as our skill increased. Given everything else, good eyesight speed in spotting the burst, and accuracy in judging distance, are the attributes essential to a good artillery pilot.

All ready to go we took off, climbed to 1,000 feet and as we circled the field the observer would release the aerial which trailed behind the plane for 200/300 feet. Then we'd check the radio with the operator down below who'd signal the O.K. or otherwise. As this was going on we were also giving the engine its final check. Government designed, as to be expected it was a constant source of worry, unreliable and with a habit of cutting out soon after leaving the ground so the final check within gliding distance of the field was absolutely essential.

Our next move was to the battery we were to direct. Arriving over them we would signal down our number for identification – mine was 7 – until it had to be changed and then circle until through the medium of canvas strips, we received the signal they were ready. Two-way radio between plane and ground had not been introduced.

Lining up the target was the next job, but before flying over to orient it with our maps we searched the skies far and wide for signs of enemy aircraft as their presence or otherwise determined our course of action. The signs being favorable, we made a bee-line for the target and with it well in view tapped out the signal to fire. As the shell burst we flashed back distance and angle and proceeded to range the next gun, and then the next until all four were ranged. With a possible 100/200 rounds left with which to wipe out the enemy battery or strong point, we gave the signal to go on firing in their own time and turned to the next job. With at least two hours more gas we had the option of directing another battery, patrolling up and down the front spotting for enemy artillery, or photograph previous shoots or anything else headquarters required. But, periodically, we would check the battery previously ranged, and if they were shooting wild, we stopped the firing and re-ranged the guns.

In the meantime, enemy aircraft may not be in the vicinity but "Archie" would be banging away like a raging madman throwing up a barrage of black smoke heavily laced with shrapnel. A few chunks may tear through the fabric and bits of metal shower the cockpits which concerned us little, but when we heard the woof and the plane began to rock, it was becoming a little too risky, so we skidded and twisted our way out of trouble and continued on ranging and expecting the worst that was yet to come.

This intermittent firing would attract enemy aircraft if they were within sight of the barrage and at any time we knew they would be on our tail ready to chalk up another R.E. 8. Sometimes they fell into a trap as they came in for the kill. High above some of our own fighters would be sitting, and as the enemy closed in down they would come, the dogfight would be on, and we would be left alone to watch a scene. Within a few minutes, it was all over and both sides pulled out and we would continue on with our work and the battery that had stopped to watch would again go into action.

Our fighters were spread pretty thin, and always on the offensive, they were generally deep into German territory and had few machines available for escort duty. So we worked alone, and often times unwittingly acted as decoys as our fighters changed their tactics, hovered out of sight high above us waiting for the Germans to be attracted by the tempting bait of the R.E. 8.

With batteries of widely different calibers it was possible to direct two batteries simultaneously due to the time element that varied between gun loadings and setting. Furthermore, as a pilot gained experience and versatility, he was able to range and keep on target three batteries before his gas ran low and it was time to leave the line. This would average around three hours.

We would notify the batteries and then, on our way home, would come down in a dive, jazz up the battery or batteries for a minute or two, to the great joy of the lads serving the guns, and with everybody waving like mad we'd toot away on the Klaxon happy as hell to be still around and make for the field.

On landing, we would taxi up to the hangar to be met by our mechanics smiling happy that once again we had outwitted the Hun. Before making out the report I discussed the performance of engine and plane with them, checked for holes and tears from shrapnel and then into the office to get on with the paper work. This completed, I would take it over to the C.O., Major Tyson – the best man I ever served under – where we discussed the flight and what had occurred. He had been an outstanding pilot and artillery man and wanted his reports firsthand and not through some subordinate as was the prevailing method with those of the old army school.

Casualties were heavy and the squadron was rarely up to strength. So, if there was time before dark, we would be up again for another three hours of artillery work, reconnaissance, or whatever was of importance. The men who didn't return, and also the hours we were unable to fly due to wretched weather, left so much time to be caught up, and so much unfinished work, that we had to fly from dawn to dusk to accomplish what we had to do. Also, one must not forget night bombing which was merely incidental and as effective as a flea bit on an elephant's hide in its tactical or strategic value. However, it all helped to increase the tonnage of bombs dropped about which headquarters loved to brag.

We flew in all weather, except in blinding fog, and even that was attempted when we were sent out to support the tanks that were attacking in mass at the Battle of Cambrai in November/December 1917.[1] Although many took off, none made it as scores of pilots and observers were killed or badly injured. Altogether, it was an incredible fiasco. Drifting fog, however, was no deterrent. Although visibility was what may be termed "in and out," a "bad weather pilot" considered it no worse than a very low overcast of limited visibility. Many pilots never could adapt themselves to it but they went in anyhow. Thunderstorms were admittedly dangerous but there was always the chance they veered, and so we took a try at it. When rain or snow was medium clear at least one could see the ground and even do a "shoot" or contact patrol as at Vimy Ridge. If it got too heavy, the wind shield offered a measure of protection with the probabilities it might ease up. The slow steady drizzle, so prevalent, had only a minor effect that interfered little with "shoots" and reconnaissance.

It was tough going but it made good pilots of those who could take it. Bad weather had its advantages. It kept the fighter planes on the ground enabling us to continue with our various tasks with only "Archie" to worry us, and of all things, the sausage balloon. Those long bags of gas held captive by a steel cable at altitudes of 2,000 feet or more were known as observation balloons. I refer to our own and not those of the enemy. These monstrous shapes were suspended at irregular intervals along the front from the channel to the borders of Switzerland. Ostensibly their purpose was to watch for activity behind the German lines, artillery direction, and observation for enemy batteries.

Before artillery direction and reconnaissance became the prerogative of the vastly more mobile planes of the Corps Squadrons, the balloons served a useful purpose, but by 1916 they were completely obsolete and became a distinct hazard to the airman. In the belief that they had a definite effect upon the morale of the enemy, many balloon sections sent them up oftentimes without observers. The cables were impossible to see in the drifting fog or low overcast and periodically some plane would run afoul of one and that would be the end of one or two airmen and a plane. We protested but without effect.

Although well defended from below by machine gunners and occasionally anti-aircraft batteries from attack by the fast maneuvering fighter planes, the balloons offered a splendid target and were shot down with increasing regularity as aircraft improved in speed and performance. Unlike the unfortunate pilots and their observers, those observers in the balloons carried parachutes which were not packed but open and ready to go. So, over the side they jumped to land safely on the ground below, that is if the pilot of the attacking plane didn't follow them down and shoot the hell out of them as the Germans often did. There was a loud outcry against this lack of sportsmanship which was a ludicrous attitude to say the least, and I'm happy to say the R.F.C. did not support the notion. There is only one rule in warfare and that is to kill in any possible manner one's enemy and put him out of business.

Incidentally, the officers and men of the balloon corps were known as "Balloonatics." Unfortunately, we had to suffer them until the end of the war when they became as extinct as the Dodo bird. But often as we saw them shot down we had little sympathy figuring as we did that another hazard, if only

Above: Member of Balloon Corps parachuting from observation balloon. Courtesy of the R. Duncan Collection at Southern Illinois University at Carbondale.

temporary, had been removed from our path.

We ranged every caliber of gun from the 18-pounder field gun that furnished the "creeping barrage," the heavy howitzers, to the long-range 12-inch naval guns that traveled up and down the front on railroad mobile units, one of the few guns that moved under traction. For power, the gunners relied chiefly on horses and mules who died by the thousands from sheer exhaustion as they were driven until they dropped in their tracks when the Army began to move forward.

In a "push" it was the artillery battery that took it on the chin. At the Arras show thousands of guns went into action. They stood in long rows and in depth. Often only a few yards separated one battery from another as their crews served the guns, keeping the men covered who went over the top. They were out in the open and often received as much as they gave. Unlike the infantry who were on the move and able to duck and dodge or take cover as they advanced, the gunners had to stand by their guns, maintain the barrage and the concentrated fire upon the enemy artillery who retaliated with equal vigor. Both sides suffered heavy losses in men and guns as the infantry were mowed down by machine guns and the gunners were blown to hell by shell fire.

Batteries were completely destroyed as the guns were tossed on one side or blown to pieces, ammunition dumps went up in flames and the ground literally burned from the intensity of the bombardment. Brave men, these gunners, as we looked down at the inferno below that continued on until the tide of battle turned.

The artillery was the strong right arm of the flying corps, the most effective weapon that we possessed with which to destroy the enemy and pave the way for a successful attack. Each recognized this and the capabilities and limitations of the other and worked as a team in full cooperation as we strove to curb the power of the opposing gunners.

Prompted by Baron von Richthofen who flew for several months as an artillery pilot at the beginning of his career and who was aware of the tremendous influence exerted by men experienced and skilled in Corps Squadron work, the artillery pilots were subject to special attention by the fighter planes of the German Air Force. Through March and April, 1917, losses were exceptionally heavy. Noteworthy was the high percentage of top artillery pilots. No sooner did they begin a "shoot" – and it was only on a "shoot" when this occurred – that they would be attacked by swarms of enemy fighters. Down they went in flames or were shot to pieces. Finally, intelligence received the information that the Germans had devised a system which enabled them to intercept the number of the pilot when checking in the battery. From the records they compiled they quickly discovered the qualifications of the various artillery men and concentrated upon the best.

We countered this the only possible way by changing the numbers so that no pilot ever used the same one twice. But the damage had been costly and during this period of concentrated effort over half the R.F.C. was wiped out with 85% of the casualties being Corps Squadron men – pilot and observer. These antiquated B.E. 2c's and lumbering R.E. 8's were von Richthofen's special meat – "cold meat" as he so contemptuously dubbed them. His records – what there is available – indicate that of his 82 victories, 69 were two-seaters of the above types. Our fortunes were indeed low as the Huns dominated the skies. As the "circuses" (*Jagdgeschwader*) increased in numbers, the artillery pilots were doomed.

Endnotes

1 The Battle of Cambrai was one of the deadliest Canadian operations of the war. More than 30,000 Canadians were killed during the six-day battle (November 20 to December 7, 1917) for the canal and marked the first large-scale use of Mark IV tanks in warfare, thus breaking the trench deadlock of the previous three years. Overall, however, little territory was gained from the operation.

Chapter 12: Bombs, Mail, and Spies

We only bombed on moonlight nights since without the aid of the moon it would have been impossible to pick out our targets – hutments, dumps, strong points, etc. Towns and cities were too deep in enemy territory and well beyond the limited range of our planes – three hours at sixty 60 miles an hour. Furthermore, our bombs were, comparatively speaking, pip-squeaks and only effective when they hit the bull's eye. A course was steered by landmarks or compass. The former method had its merits in fine, clear weather only. The latter, a dead reckoning, followed a plotted course aided by compass, watch, and air speed indicator – the only sure way to navigate.

All pilots had to pass the test in navigation and were supposed to fly by it, but far too many placed little faith in the compass, preferring to operate by checking landmarks from their maps as they flew their course. They were the ones who were so frequently lost, and as often as not, killed. Compasses were extremely reliable if checked regularly and corrected. This had to be done by the pilot. It was simplicity itself, but a job disliked by many. So it wasn't done.

For night assignments, we flew alone. There was no need for an observer or aerial gunner as the fighter planes kept to the ground at night. They were keen to go but it was risking the life of a well-trained man unnecessarily. The pilot released the bombs himself either independently or in one cluster.

My first bombing expedition was within three weeks of my arrival at the Corps Squadron by which time I had a total of roughly forty hours of flying which included flight time under training. There were three of us making our first night flights without having had an opportunity to practice a few approaches and landings on the field. The target on my first flight was a supply dump located in the ruins of the village of Serre about six miles behind the German lines.[1] It was a cold winter's night in January 1917 with the moon in the full. Perfect for an inexperienced pilot. I had difficulty in finding the spot as nothing but the outlines of the village streets and foundations of the houses remained. It was flattened in the Somme battles of 1916. However, I was able to get it in focus after an hour's intensive search, spotted the dump and dropped my four 20-pound bombs. I circled the area a couple of times and then, fearful my engine might conk, I headed back to the field. What damage was done I never did find out.

It was a wonderful evening, visibility was perfect, the air smooth and the engine was literally purring from the perfect mixture of cold night air and gas. So I sat back at a thousand feet, perfectly at ease, watching the panorama of war stretching north and south as far as the eye could see. It was a thrilling and marvelous display. Guns flashing, shells bursting and lighting up the ground and then back to blackness; parachute flares of the Germans flood-lighting the trenches and barbed wire of "No Man's Land," along with the many multi-colored Very lights[2] of the British. Then down the length of the trenches seeking out and killing anything that moved. It was like a steel city with its myriads of furnaces shooting tongues of flame into the skies, endlessly lighting up and dying down throughout the night.

After a perfect landing, virtually in the dark, I lost all fear of night flying as that had really been my principle worry. No fighter planes patrolled the skies, our only interference and concern being "Archie." And, without searchlights, "Archie" was useless. Batteries equipped with them, once they held us in the beam, tossed up a weird assembly of phosphorous balls strung together which we dubbed "Flaming Onions." Barrages of them were thrown up at bombers when they flew within range. The theory was that once they became entangled in the rigging, the plane would catch fire. Whether this weird assembly met with success, I never did hear. It was a German invention and a most fantastic device but utterly useless in rough weather.

The searchlights actually were a boomerang. Once their beams focused on a plane the pilot would point his nose down into their length and spray the immediate area below with machine gun fire. Operating from a low altitude, this proved effective as invariably the light went out and stayed that way. However, in the darkness it was impossible to see the results.

Bomb loads consisted of our 20-pound bombs or one 112-pound bomb. The smaller bombs were fitted into racks attached to the two lower wings. Each rack could be operated independently. For the 112-pound bomb, a stronger rack was suspended to the bottom of the fuselage sufficiently far back so that the bomb could clear the landing gear as it was released. When this was to be dropped, the lighter bombs were not carried, being too much weight for the plane.

When the R.E. 8 replaced the B.E. 2c, although the engine developed 60 horse-power more, the bomb load remained the same. The only difference was a 112-pound bag of sand that was dumped in the rear cockpit in lieu of the observer, as without this extraordinary ballast the plane would go into a nosedive. An aerodynamic error that had been fatal to a number of pilots before it was discovered. It was described as a "moveable center of gravity" and was never corrected so the forgetful pilot, as he climbed to a certain altitude and throttled back to level off, would go into a steep dive from which there was no recovery. I saw this happen to a pilot who had been posted to my flight. He had never flown an R.E. 8 so I told the Flight Sergeant to have a ship ready for him to take up and practice. I then walked over to the orderly room, completed my business, and as I was walking back to the hangar saw the chap take off and level off at about 1,000 feet. The machine hesitated for a moment and then went into a steep nose dive, hit the ground, and spread itself over the landscape. I rushed over to the crash at the far end of the field, along with a crowd of mechanics and poked around to fine the cause. The pilot was dead with a broken back from the terrific contusion. The reason for the accident was a careless mechanic who had forgotten to put the necessary ballast in the rear cockpit. The unfortunate pilot was only one of several hundred killed by this aerodynamic nightmare.

Our night raids were carried out with the utmost secrecy. The objective was not given to us until shortly before leaving the ground but allowing sufficient time for us to pinpoint the target and plot our course. Throughout the bombing period, generally between midnight and 3:00 a.m., the field would be in complete darkness except for the few minutes it took the plane to land.

Navigation lights were installed on the wing tips and tail fin as is the custom today. These were the only lights on the plane except for tiny shaded lights on the instrument panel which could be switched off until we crossed our own line again on the return trip, if we made it. Starting with a Lewis gun on the B.E.2c and a Vicker's front gun on the R.E. 8, standard equipment included a Very pistol, flares, Klaxon horn, maps, and writing pad. Other essentials included a radio, a lead-weighted message bag, and streamers.

On the return trip, when crossing the line, the more reckless of us, or those who hated a German's guts, would shoot up the enemy trenches and then buzz down insulting messages to the angry line below on the Klaxon horn. This pleased our own lads and we had them in mind when we did it. We were the U.S.O. shows of that time and would do anything to relieve the monotony of those living in the dreadful trenches. It gave us a good laugh too, and those crazy Germans, seething with rage, would toss up everything they had – magnesium flares, pistol flares, machine gun fire and even the rifles would empty into the blackness above. As the infantry often told us, we certainly had those Heinies tearing their hair out.

As we approached within landing distance of the field, we would signal our return with a certain colored Very light and cruise around until the landing lights were lit and the signal to land given. There were no flood lights or field lights of any description, nothing but an impromptu arrangement of petrol cans laid out in the form of an "L" outlining the landing strip. The approach was toward the bottom of the "L." The cans were filled with gas-soaked rags which were lit prior to the signal for the plane to come on in and were doused once the plane was on its way to the hanger. Even with the utmost precautions, however, enemy planes would occasionally spot a field and before long their daylight bombers would be over and give it a good pasting. This happened to us about 1:00 p.m. when we'd moved up from Aubigmy on the Arras front to the field adjacent to Mount St. Eloi. The mess emptied like a flash with everyone diving for the nearest hole or ditch but no one was hurt and the damage was slight.

Information relative to the location of the various night targets we bombed was secured later through photographs, reconnaissance and intelligence. Also, the Allies had a network of spies operating in German territory, many of whom moved back and forth by devious channels furnishing our intelligence with the information they had gathered. Once a report dealt with a large group of hutments the Germans were using for rest billets a few miles west of Dovai and north of the river Scarpe. An easy target to line up as the river stood out like a silver streak in the moonlight. Also, the report gave the exact bearing of the place and adjacent landmarks. We were now back at the Arras front and this was a week or so after the battle had been called to a halt in the early part of May. We had recently been re-equipped with the R.E. 8, a faster job than the B.E. 2c and with greatly improved armament but we still carried the same amount of bombs. For this affair I carried four 20-pounders to give me a better spread.

For the first time I was to carry a passenger, a staff intelligence officer who wanted the thrill of a bombing inspection along with that of night flying. And, for no other reason, I was happy to have him come along, especially as he had had some experience with machine guns. The observer's cockpit was now in the back, equipped with two guns which could be swung in any direction so I could make good use of him after the raid. Incidentally, he was a cavalry man, a major from the Dragoon Guards and was later attached to the R.F.C.

We had no difficulty finding the hutments. It was about 2:00 a.m. with everyone tucked away in their bunks and dreaming or snoring when I let my bombs go, and with effect. One must have hit a fuel tank because I'd hardly got clear before flames were shooting up. I circled once at a safe distance and then, satisfied that we'd one a good job, I headed back. As we drew closer to the line we spotted a column of vehicles standing out clearly in the moonlight, heading in the direction of the line so we dived down on men, horses, and carts, giving them the business from our three guns. In a matter of seconds, the road was a mass of plunging horses, jammed up vehicles and with those that could, tearing across the open fields in all directions. I felt sorry for the poor animals, many of whom were undoubtedly killed or badly wounded, more so than for the men who could take care of themselves. In the excitement, the major tossed an empty ammunition drum overboard which, caught in the slipstream, tore through the rudder ripping a good-sized hole in the fabric. However, it had little effect on the controls and we made it home without incident.

This turned out to be the last night expedition I ever made with my gallant friend, the Major, but we did more damage than we first thought. Information came through that the hutments were destroyed by fire, casualties were heavy, including 100 men killed, and all with four pip-squeak bombs. As they truly say, "It isn't the size of the bomb but how you lay it."

Occasionally on a bombing expedition we would carry a small sack of mail which could be squeezed in alongside the

Right: Figure 40. Ernst Handley-Page O/100 twin-engine British bomber. The underside of the wings and struts were camouflaged in black for night bombing. Courtesy of Aeronaut.

pilot. It contained letters from prisoners of war to friends and relatives in Germany. They were unsealed so we read a number of them in which the writers raved about the excellent treatment and food they were receiving. After we glanced through a number of letters that were similar in the handwriting, the contents indicated they were form letters run off on a copying machine. But the addresses were correct and the signatures varied.

This was excellent propaganda and feared by the Germans who resorted to extreme methods to stamp it out. Orders were issued stating that anyone found in possession of such material would be shot out of hand. Furthermore, any pilot captured along with the evidence that he had been dumping letters around the countryside would receive the same treatment. The Germans were thorough, ruthless, and excellent fighting men. Their fault was too much organization because it developed a distinct lack of initiative. They could have handled the letter situation much better instead of indulging in their usual practice of trying to drive the fear of God into the troops and particularly the R.F.C.

We did not fly over in mass formations. Planes took off at spaced intervals, dropped their bombs, and returned along a different course supposedly to avoid a collision in the dark which could easily happen according to the law of averages.

This phase of our work was by no means popular with a percentage of pilots. Some had an inherent fear of night flying for one reason or another and would drop their bombs on some farmer's field on our side of the line, stall around to fill in time, and then return. When suspicion was aroused, the next time he left on a raid he would be followed and if caught, his career in the R.F.C. would be over. There was no court-martial. If a regimental officer, he was returned to his regiment, and when such was not the case, he ended up as a general rTo most pilots, however, no matter how useless we considered tossing those pip-squeak bombs around, it was a job to be done and to be done right. To say the least, it displayed a lack of relative values in sending a highly skilled pilot and an expensive plane, especially at a time when the shortage of both was so great, to drop four miserable fire crackers on a target of more than dubious value. We freely expressed our opinion but headquarters paid little attention to our gripes. They dealt in tonnage – 100 pip-squeaks made a ton – because certain Army commanders were more concerned with the tonnage of bombs dropped than the actual damage inflicted. If this was greater over a given period of time than others, it became a source of personal satisfaction and another step up the ladder.

The two-seat German bomber, the Albatros C III was a day and night bomber like the type in which von Richthofen began his career and dropped its bombs from a height of 8/10,000 feet. Daylight bombing was a nerve wracking job, especially without an escort as these fellows discovered. On their way back they ran into a fight of S.E. 5s who were waiting for them and in the dog-fight that followed two were brought down on our side of the line without too much damage.

By May 1918, the Independent Air Force was organized and put in operation. It was for bombing purposes only and employed in destroying industrial centers such as Trier, Frankfort, etc. They operated under their own commander-in-chief, General Trenchard. In these operations the Handley Page, a twin-engine bomber played a prominent part. Capable of carrying 16-112-pound bombs or 8-250 pounders, they proved most effective and inflicted appreciable damage on the enemy. The day of the pip-squeaks was over.

Spies

On both sides of the line spies were everywhere. When they were caught they were shot without ceremony or undue delay.

The great majority were native French or Belgians giving aid and assistance to whichever side they believed would win. Their pay was miserable and the risks far greater. But the majority of those working for the Allies were truly patriotic and among the bravest in the war.

Many of those who had been working with us behind the German line were enthusiastic Frenchmen drawn from all classes and composed of people native to the territory from which they operated. They were well organized, divided into groups under leaders many of whom passed back and forth from one side to the other with essential information required by the Allies. Some made their way through neutral countries, others even crawled out between the lines and through the wire. Theirs was the long shot and many died.

However, it was harder to get in than to get out, particularly for the master spies controlling the operations covering territory adjacent to the battle areas. To handle the problem of getting these men to the scene of activities with the utmost speed and with reasonable safety another responsibility was added to the manifold duties of the Corps Squadron pilot. This is how it was accomplished.

The two main essentials to insure success were a pilot with exceptional guts and initiative, experienced in night flying, and a plane with an engine of proven reliability. Engines varied like mules in contrariness and refusal to operate with certainty except under ideal weather conditions. Air-cooled motors such as we were saddled with were particularly difficult in cold weather. However, thanks to the genius of an engine mechanic expert in ignition and carburation, I had a motor that would start, regardless of how cold the weather was, on the first swing of the prop. This was abetted by placing an ingeniously devised heater under the cowling close to the carburetor. And that is how and why my plane made a trip across the line, dropped a spy, and returned safely to its hangar.

The utmost secrecy was observed when a job of this nature was to be carried out. However, it was common knowledge that a plane was to be readied for a special night flight but the purpose of this was known only by the C.O. and Intelligence at Army Headquarters. Even the pilot was unaware of his destination or mission until a few minutes before takeoff.

"Shorty" Taylor was designated to fly my plane and he was aptly named. In build and appearance, he was a typical jockey, 5 feet 3 inches in height and weighed 110 pounds. Whiskey, old ale, and the easy going life of a pilot had increased his girth and weight by 20 pounds. But he was the ideal man for the job, physically coupled with all the necessary attributes of the night flying pilot.

One evening when the moon was on the wane but visibility was sufficient, an order for takeoff at 2:15 a.m. came through. It was a cold winter's night towards the end of January and the weather was clear. Nothing was left to chance and at 2:00 a.m. "Shorty" was sitting in the cockpit with the engine warmed up and the prop slowly ticking over when out of the dark a car emerged, pulled up alongside the plane and two men got out. One was the passenger and the other a captain from Intelligence. "Shorty" climbed out of his seat and hopped to the ground. A quick handshake and then, as the three men huddled over a map, he discovered his passenger was a major link in the chain of intelligence, or spying, behind the German lines and his job was to get him to his territory.

With the course plotted out and the destination pinpointed to "Shorty's" satisfaction, both men climbed into the plane and then he realized why the weight of the pilot had been a matter of concern. His friend, the spy, was a huge fellow, well over six feet and heavily built. But until the old B.E. 2c with its 90 horse-power engine finally left the ground, he didn't realize his weight which he estimated to be at least 200 pounds. However, loaded to capacity as it was, the engine and plane took it in stride.

Climbing to the highest possible altitude from which he could spot his landing site, "Shorty" crossed the line with all lights off and quickly arrived at his destination. This was high-lighted by a large oblong woods running parallel to which was a cut over a grain field free of obstructions and offering ample room for landing and takeoff. Incidentally, it is far easier to get into a landing area than to get out. He circled the area first, picked his spot for landing and then pulled back the throttle and began the lone but noiseless glide to earth. As quickly as the plane came to a stop his passenger scrambled out, muttered a hasty "bon chance" and disappeared into the gloom. "Shorty" had made it. But he was twenty miles behind the German line, and although there was no indication of any enemy installations or troops within the area, he was in one hell of a situation. His engine was "dead" – my engine that never failed!

For a moment, as he told me later, he sat there like a man who'd been knocked out cold, and then as his brain began to work and reason returned, he realized how stupid he had been and the cause of the engine going "dead" – it was all chocked up and cold. As he glided down he forgot to give the engine the throttle periodically in order to clear the surplus gas being sucked into the cylinders as the prop turned over. In fact, he made a dead stick landing but in the excitement of unloading his passenger, hadn't realized it.

Without further ado, he switched off, hopped out of the cockpit, pulled the prop backwards several turns to clear the cylinders and then went back again to set the throttle for sucking in fresh gas. As he did so, he remembered the wheel chocks without which, once the engine started, the plane might get out of control. There were no wheel brakes in those days. So he walked over to the nearest woods for something to use as blocks and by sheer luck came across a stack of short-pit props. Added to all this was the worry that the Heinies could be scouting around looking for him. If caught, he would be shot on the spot. This was the justice meted out to all spies or those who assisted them.

Placing the log in front of the wheels, he pulled the prop

through several revolutions by which time he was sweating freely, encumbered as he was with a heavy leather wool lined flying coat. So, as he returned to the cockpit to switch on and set the throttle, he was in one hell of a hurry and tossed the coat on the ground alongside the landing gear, came back and grabbed hold of one blade of this four-bladed "wing and a prayer," put all he had into one terrific swing and by the grace of God it kicked off. In a split second he was round the wing tip and into the cockpit where he sat for the longest seconds of his life warming up the engine. He revved it up slowly, checked the rpm's for power, pulled back the throttle, then pushed it forward again for takeoff. The tail came up and the plane almost tilted on its nose as the log held the wheels fast. Again "Shorty" shot out of the cockpit, pulled the log out of the way and made it back as the plane slowly inched its way forward. Now he really was in a sweat. As a matter of fact, he began sweating all over again as he told the story.

However, without further incident he took off with no one in sight and within thirty minutes was back at the field. After checking in, he tossed off a couple of "Whisky Macs" chasing them with a bottle of Bass ale and went to a well-earned rest.

What happened to the spy nobody knows but it proved rather costly to "Shorty." The leather coat he discarded was not a government issue and in the excitement and urgency of the moment had forgotten all about the coat until he was straightened out and safely on his way from a bullet in the head. No doubt some worthy French farmer had a much warmer winter than anticipated.

Endnotes

1 Serre was one of many fortified villages held by the Germans at the start of the Battle of the Somme in 1916.

2 Very lights were flares fired from a pistol and used for signaling and lighting at night. The Germans used flares shot into the air and suspended from a small parachute.

Above: Friedrichshafen G.IIIa night bomber. The single fin and rudder of the G.III were replaced by dual fins and rudders in the G.IIIa to enable better directional control in case of engine failure. German crews preferred twin-engine bombers for night bombing but experimented with single-engine night bombers. Courtesy of Aeronaut.

Above: With its long wing-span the AEG G.V was the final AEG bomber. AEG bombers were used for night bombing together with their Friedrichshafen and Gotha twin-engine counterparts; all used the 260 hp Mercedes D.IVa. Until the last months of the war the Allies mostly used single-engine pushers for night bombing. Courtesy of Aeronaut.

Chapter 13: Richthofen

Manfred von Richthofen, the legendary Red Baron, was a name to conjure with by those who flew in the First World War, and a name to be feared by the lone artillery pilot to whom he gave special attention. By dogged perseverance – in his early training, he was a hopeless student and cracked up repeatedly – and a fierce ambition with a lust to kill, he was determined to strike terror in the hearts of his enemy. And this he did to many. To others he was the foe they longed to meet, to challenge, and to send to his death. He avoided none and was utterly fearless in the cold, detached manner of the Prussian officer.

He introduced mass formation and mass attack and put his theories to work once he had command of the original "circus," *Jagdstaffel* No. 2. Moving across the skies, often in a long line of planes flying side by side, once they met an enemy aircraft they would peel off, one behind the other, pouring heavy and concentrated fire into their victim until down he went down in flames.

Whatever may be controversial about the Red Baron and his cold-blooded Prussian attitude toward friend or foe, it cannot be said he was a shrinking violet who took cover behind his men, although he did sit back and pick off the stragglers. The brilliant scarlet of his plane dispelled that theory and the bizarre system of painting adorning the planes of his "circus." Intended to instill a measure of fear, it failed in its intent. But it brought all the color of the "circus" to those drab grey skies strongly appreciated by the R.F.C. and making, as it did, identification at a distance simple. The background upon which these weird designs was splashed was a brilliant scarlet. His own plane remained a vivid scarlet, unadorned other than by the insignia of the Iron Cross. Instead he flaunted his contempt for the R.F.C. and along with it a challenge as he led his "circus" across the skies – a study in scarlet that earned for him the sobriquet of the Red Baron. From September 1917 until his death in April 1918, he flew a specially constructed Fokker Triplane even though, after less than five months in the air, this model washed out and all squadrons were re-equipped with the Albatros D III.

Subsequently, this bizarre system of painting as opposed to camouflage, was adopted by all the Jagdstaffeln, with each "circus" varying in size from fourteen to thirty planes. However, unless one spotted the red plane of the Baron during a dogfight, the identity of the his "circus" was left in doubt.

One time when I was filling in the gaps after a mapping flight, according to observers from the ground, I met him and roughly twenty planes traveling in a long line, two by two, just north of Arras in August, 1917. Down they came, peeling off one behind the other on my tail and blind side with their two guns streaking the skies with racers. But Jock, my observer, a fine gunner, blazed away and held them off as I twisted and turned, edging towards a huge cumulus cloud floating majestically across the sky. By now they were all around me below and above. I could hear the sharp crack of bullets ripping the fabric, any minute expecting to get it but thinking only of that huge white refuge which I finally made and dived into with wide open throttle, nose straight down with not a second to spare. That was the closest call I ever had as three or four planes came at me from all angles.

I never realized the Baron was there but the men below spotted him as we were comparatively low, at an altitude of not more than 4,000 feet. I came out of the cloud at about 2,000 feet, straightened out, and scanned the skies and looking for the Huns and was ready to climb back in if necessary. They were in formation again, probably marking me up as another victory and heading north seeking another easy mark.

Richthofen had established himself by now and was recognized not only as a man of many victories but also as a superb tactician in whose judgment the High Command had implicit faith. He was free to move as he saw fit and was not confined to any sector of the front, but always to be found wherever preparations for a "push" were in progress and dealing death and destruction to the unfortunate pilots plugging away in their R.E. 8's. Then, like a guerilla fighter, he would move on to new pastures as our fighters vainly scoured the skies looking for the man who wasn't there.

He picked his men carefully like his predecessor, Boelcke,[1] visiting other fighter squadrons and selecting those who'd proved their worth. Among them was his brother, Lothar, who also chalked up a notable list of victories.[2] In character and disposition, Lothar was the direct opposite of Manfred, having a friendly carefree manner, always ready to enter into the celebrations and pleasures of the mess. Here Manfred Richthofen is standing by the cockpit of his famous red triplane in leather jacket and carrying a walking stick, looking fit and ready to go. Lothar is standing alongside him overshadowed by his brilliant brother.

I'm discussing von Richthofen at length because he was the nemesis of the artillery pilot in addition to being the outstanding ace of all in World War I. The nearest any pilot came to his total of eighty-two victories was the Frenchman, René Fonck, with a score of seventy-five.[3] Richthofen established the German strategy of defense as opposed to the offensive spirit of the British. As he so aptly put it: "The customer should come to the store and not the store to the customer."

The Baron was not popular with his own men being a typical cavalry officer of the Prussian school, a strict disciplinarian, cold in his attitude towards others, including his brother officers, and with a profound contempt for those of lesser degree. But he was a first-class fighter pilot, cautious, cunning, and brave who, once he succeeded to his own Jagdstaffel II, which was organized by Oswald Boelcke,

Above: Lothar von Richthofen (seated on ground) with other members of *Jasta* 11. His brother, Manfred, aka the Red Baron, is in the cockpit of this Albatros D.III, April 23, 1917. Courtesy of Aeronaut.

the originator of the "circus" method of mass formation, improved it still further by his tactics and the ruthless discipline he exercised. He was a killer, without sympathy for his enemy or his own.

To Richthofen, flying was not the sport and adventure it was to so many pilots. To him it was a means to an end, the satisfaction of meeting an opponent face to face with the assurance of a superman, a means of quenching his thirst for blood. He was fond of describing the agonies of his victims as he followed them down to gloat as they writhed in agony from the heat of the flames, fanned by the draft of the falling plane as it burnt to a cinder.

He was a man who calculated his opponents according to their worth and he considered the most dangerous of all to be the artillery pilot. As such he began his own career in the air force as an artillery pilot and so knew from experience the deadly character of their work. The fighter pilot was a destroyer of individuals, but the artillery pilot decided the fate of hundreds and into the thousands.

His reasoning was sound and so they became his prey, not so much because of the comparative ease with which they were shot down but because of the destruction they wrought. And it can truly be said he never refused combat and fought to the finish with any fighter pilot who sought him out. Our finest did. His plane stood out boldly, a study in scarlet, as if to say: "Here I am, come and get me."

When the war broke out, Richthofen was an officer in a regiment of Uhlans, a type of light cavalry primarily armed with a lance. But, convinced the days of the cavalry were over, he applied for an appointment to the German Air Force. Trained as an observer, he was sent to the Eastern Front against the Russians. Eventually, he qualified for his pilot's wings and arrived at the Western Front in March 1916 where he was posted to a two-seater Corps Squadron equipped with Albatros C III. This biplane carried out the same duties as the B.E. 2c and the R. E. 8 but was far superior due to its excellent engine, the Mercedes 170 horse-power. Like the British Rolls-Royce, it never let the pilot down.

However, he was determined to become a fighter pilot, but it wasn't until August of the same year his ambition was finally realized when he became a member of Oswald Boelcke's squadron of combat planes. The next month Jagdstaffel II was formed by Boelcke and the first formation of fighter planes in mass moved across the skies – the beginning of a flock of "flying circuses."

In October 1916, Boelcke was killed after scraping wings with one of his own men in a dogfight. Not in mortal combat with a worthy challenger. He died ingloriously as did so

Left: Ernst Udet and his Albatros D.V. Member of famous Tango Circus—Blue Tails. Here Udet was photographed as commanding officer of *Jasta* 37. He finished the war with 62 confirmed victories, the second highest-scoring German ace after the Red Baron, who had 80 confirmed victories. Lothar von Richthofen, the Red Baron's younger brother, scored 40 confirmed victories. Courtesy of Aeronaut.

many of the great aces – not in the fierceness of combat as the machine guns blazed, but in a manner so unexpected and so simple it could only be declared stupid. Eventually Richthofen took over command of the *Jagdstaffel* and made it the most famous of all the numerous German "circuses." His greatest personal victory, his eleventh, occurred in November 1916 when he shot down Major George Hawker V.C., D.S.O.,[4] the leading ace in the R.F.C. at the time, after a terrific fight lasting forty-five minutes as they sparred, pulled out and came back again in a battle to the finish. Ending a few hundred feet above the ground, Hawker fell to his death behind the German lines and was buried the next day with full military honors.

The Red Baron, well in the saddle, was striving for a mark of 100 and became ruthless in his claims, domineering in his attitude towards those with whom he flew, and merciless to those who crossed him up. The obnoxious traits of the Prussian officer grew in proportion to his victories and honors. He was never known to express sorrow at the loss of a pilot and grudgingly, and only because of necessity, bestowed the benefits of his experience and skill upon new arrivals. Little sorrow was expressed at his passing as in their hearts they were glad to see him go. Hermann Göring, not then the fat mass of blubber who, for a time, was Hitler's right hand, but a gay slim young pilot with many victories to his credit, was the last commander of the famous Richthofen "flying circus."

On April 21, 1918, Richthofen and his "circus" clashed with a strong formation, three flights of Camels from Squadron 209, one of which was led by Captain Roy Brown. In Captain Brown's flight was a young pilot recently arrived from England. The weather was perfect and visibility excellent. As the dogfight settled down into a series of individual scraps, with planes darting all over the place and sparring for an opening, tracer bullets streaking a phosphorous path through the skies, Brown cast an eye around to see how his young pilot was doing, a chap by the name of May from Edmonton, Canada. He noticed that May had broken away from the pack and was heading back across the line obviously in serious trouble. And right behind him, in line with his tail, was a red triplane, closing in fast. Like a flash, Brown swung over, caught up with the red machine, lined him in his sights and gave him a quick burst. Down went the plane into a steep dive and made a safe but rough landing behind the Australian lines without firing a single shot. With a sure kill in his sights and neglecting to give that final glance to the rear, the Hun had been trapped like the veriest fledgling. That was the end of the scrap and Brown and the rest returned to their field unaware of the sensational news awaiting them.

As the plane rolled to a stop, the Aussies rushed over and there sat the pilot still holding the stick, dead. Like the great pilot he was, he'd stayed with the ship to the last bringing it safely back to earth. They lifted him out, searched his pockets for identification, and discovered to their amazement the dead aviator was the Red Baron, the man who had become a legend even before he died. One bullet was enough, through the chest. They buried him with full military honors in the churchyard of a village a few miles away with the Aussies furnishing the firing party.

Captain Brown, a first-class fighter pilot – 12 victories – had no inkling of the identity of his victim and no man was more amazed when he heard it and at the ease with which he had polished him off. His was the outstanding victor of all the battles in the air in the War and one that every fighter would have given his right arm to accomplish. And no group of men were more delighted when they heard the news than the Corps Squadron pilots. Richthofen had died like the "cold meat" he sneered at. And, in crossing the line, he broke the rule he had set for himself and his pilots – "Let the customer come to the store, not the store to the customer." Yet, despite his Prussian disposition, he was the greatest air tactician of the war and a strategist whose policies became fundamental up to and through World War II.

In April 1924, I was in Berlin where I met Ernst Udet, Germany's second ranking ace – 62 victories.[5] In the course of my stay he invited me to a dinner of former war pilots – many of whom spoke English. Naturally, Richthofen became a point of discussion and much of what I have written here is the result of what I garnered.

Endnotes

1 Oswald Boelcke (1891-1916) was credited with 40 aerial victories. He developed some of the earliest flying tactics used by the German aces and making it possible for the later achievements of Richthofen. He was killed in a mid-air collision with his good friend Erwin Böhme.

2 Lothar Siegfried von Richthofen (1894-1922) was a younger brother of Manfred von Richthofen and credited with 40 victories.

3 Colonel René Paul Fonck (1894-1953) was a French aviator who became the top Allied fighter ace with 72 solo and 3 shared victories out of 142 claims. After the war, he returned to the military and served until 1939.

4 Lanoe George Hawker (1890-1916) had seven credited victories and the third pilot to receive the Victoria Cross for gallantry.

5 Ernst Udet (1896-1941) became a squadron commander under Richthofen and later under Hermann Göring, and with 62 victories he was the highest scoring German fighter pilot to survive the war. During the 1920s and 30s he was a stunt pilot before joining the Nazi Party in 1933. He was involved in the development of the Luftwaffe and became a Luftwaffe Colonel-General (Generaloberst) during World War II. A combination of stress and a deterioration of his relationship with the Nazi Party led him to commit suicide.

Right: The Salmson 2A2 was the most common corps reconnaissance airplane used by the French and Americans in 1918. Powered by a 260 hp Salmson water-cooled radial, it had excellent performance and maneuverability. Courtesy of Aeronaut.

Chapter 14: Anthony Fokker

Supremacy of the air was juggled back and forth throughout World War I, and the decision as to which side should hold it definitely rested in the hands of the aircraft and automotive engineers when "procurement" permitted. Competition was fierce as the engineers of the combatant nations improved on the designs of the others and causing supremacy of the air to change hands. But the man who stood head and shoulders above them all was Tommy Fokker.[1] He built three series of his first plane, the Eindekker Scout, a low wing monoplane powered respectively with the 80, 100 and 130 horse-power Oberusal Rotary engines, each superior in performance and armament to the other. The No. II series was the first German plane to be fitted with a front machine gun firing through the prop in the spring of 1915. To compensate for the extra weight, the 100 horse-power Oberusal replaced the original engine. A later model, the Eindekker III, was fitted with two front guns (see facing page).

This radical improvement in fire-power enabled the Series II and III to overcome much opposition and the Germans held supremacy of the air from April 1915 until the following summer of 1916. Then came the turn of the British who hung onto it with their new models until the Albatros D.I appeared in the skies September 1916.

It was not until May 1917 that the British, with the arrival of the Sopwith Triplane, the S.E. 5, and the Bristol fighter, again secured control of the air. They cruised along in great style, knocking the German planes out of the skies but only until September of the same year.

In the meantime, Fokker went into an eclipse after the introduction of the Albatros, but he was far from out. His next model was the Fokker Triplane which went into service August 1917. Several months prior to this, a Sopwith Tripe was forced down behind the German lines and turned over to Tony for inspection with the suggestion that he endeavor to produce one that would outperform it. The R.F.C. had been very successful with this type and the Germans were impressed with its rapid rate of climb, a decided advantage in the endless struggle to get on top of another fellow's tail. The forced down plane made its last journey to a German hangar to be dismantled and broken down, section by section, for examination and study.

In performance, the Fokker Triplane was a decided improvement over its British prototype. Rate of climb was six minutes to 10,000 feet as against eleven minutes; maximum speed at 8,000 feet was 121 mph as against 116 mph at 6,5000 feet. For armament the Fokker had two Spandau machine guns, the British plane had only one – the Vicker's gun. Maneuverability at higher altitudes was far superior and its faster rate of climb gave it a considerable advantage. But in order to achieve this superiority Fokker had sacrificed structural strength. Loaded it weighed 200 pounds less than

Above: Fokker E.III 105/15 being inspected. The Fokker E series were the world's first production fighters. Courtesy Aeronaut.

Above: Oswald Boelcke in the cockpit of his Fokker E.IV. Powered by a 160 hp Oberursel U.III two-row rotary, its twin machine guns are visible. Courtesy Aeronaut.

the British plane. Cutting weight in those days was a definite determinant in the factor of safety and inherent strength of a structure. A serious error for Fokker to commit.

Airplane design then, as now, was a matter of compromise. If strength was desired, speed and maneuverability were sacrificed for weight. But when the two latter were sought, weight and wing area were sacrificed.

Fokker's Triplane did not stand up in a dive. Under stress, the wings collapsed or the fabric would be stripped from the upper wings. Consequently, the German High command stopped production, withdrew the plane from service at the end of 1917 and re-installed the Albatros D III. It was a drastic step to take for which they well deserved the highest commendation. At least they placed a higher value on the lives of their pilots and unwittingly cleared the way for a far deadlier machine with which Fokker redeemed himself, namely the Fokker D VII.

Fokker pleaded with the High Command to remodel and strengthen the D III which was comparatively simple to do and would have had only a minor influence on production. But they were adamant. However, Richthofen, for whom Fokker had a tremendous admiration, was fanatically devoted to it. By this time, he was supreme in the air, covered with honors and the pride of the German people. So he was granted permission to keep his plane provided it was rebuilt and strengthened. At the factory it was rebuilt under Tommy's individual supervision. In fact, he did much of the work himself, and on completion it went back into service where Richthofen subjected it to every test a pilot could devise. Incidentally, these violent stunts were carried out minus a parachute and so the pilot had to decide to stop at the crucial moment when the plane felt sloppy and was about ready to disintegrate. A feeling of sloppiness in the wings was a sure indication as I know only too well from experience.

When the D III first appeared on the western front in August 1917, Lieutenant Werner Voss, who was credited with forty-eight aerial victories, shot down twenty-two British planes in twenty-one days. Of these, seventeen were R.E. 8s. He was killed the following month being shot down by an unknown pilot. Voss's success was, without a doubt, due to the strong resemblance of the plane to the Sopwith Tripe. It fooled both my observer and myself, but we got away before they could get to us. Richthofen flew it to the end.

Even before the plane was washed out, the Germans were planning a new fighter that would prove a thorn in the side of all pilots on the Western Front – British, French, and Americans. The design of this plane was made a subject of competition in which would be incorporated improvements and advanced ideas of Germany's leading test and fighter pilots.

Every aircraft manufacturer was given the specifications and ordered to submit a completed plane by a certain date for tests so that they could be flown at the same time. Fokker scored decisively with a design far superior to the rest of the aircraft industry – the Fokker D VII powered with the 200 horse-power Mercedes. Production on all other models was stopped and his competitors – A.E.G. (Allgemeine Elektricitäts-Gesellschaft), Pfolz Flugzeugwerke, Albatros, Hallenstadt, Aviatic, for the rest of the war, concentrated production on this plane.

In May 1918 the Fokker D VII came into service and all fighter squadrons were equipped with it as rapidly as possible. A marked feature was its ability to maintain maximum control and maneuverability at altitudes where, because of the thin air, other aircraft were sloppy and slow on the controls. It proved to be the greatest menace of the Allied air forces

Above: Fokker D.VII 244/18 is seen here in the markings of *Lt.* Alois Heldmann. *Jasta* 10 upgraded to the BMW-powered D.VIIF in late July, so Heldmann´s aircraft could be made available to *Lt.* Lothar von Richthofen when he unexpectedly returned to the *Jagdgeschwader* on 25 July 1918. He had recovered from a broken jaw, an injury received as a consequence of a crash in a Dr.I on 13 March 1918. (Peter M. Grosz Collection/STDB)

Above: Taken at Chambry in early August, Fokker E.V 148/18 is already marked with the full *Jasta* 6 unit marking scheme of the black and white striped horizontal tail surfaces and wheel covers. The cowling has now been finished in a black and white "petal" design. The plane carried the Fokker Werknummer 2789 and was accepted at Schwerin on 13 July. The E.V was the last Fokker type to reach the front. (Peter M. Grosz Collection/STDB)

since the days of the *Eindekker* II and III. With this machine the Germans regained supremacy of the air and maintained it until the end of the war.

Fokker's final contribution was the D VIII (See facing page below) which again won out against all competitors. This was a high-wing monoplane powered with a 120 horse-power rotary Oberursal. Fortunately for the Allies, it did not appear on the scene until October 1918 and no more than forty reached the line, but in that short time before Armistice Day, it proved more deadly than the D VII. Göring, who'd succeeded to command of Richthofen's crews, shot down three Spads with it in one day, five days before the Armistice. The pilots of the R.F.C. dubbed it the "Razor Blade." A rather interesting feature is that it was a monoplane equipped with a rotary engine as was his first model, the *Eindekker*. Incidentally, the Oberusal engine was the La Rhone which was manufactured in Germany under license from its parent French company.

Thanks to Fokker, the Germans were the first to adopt the front gun firing through the prop as standard equipment. But the first pilot to fly a plane so equipped was the famous pre-war pilot, Roland Garros, a member of the French Air Force.[2]

The gun was designed by Eugene Gilbert, a Frenchman, and first put to use for aerial combat by Garros who redesigned and vastly improved the original model. Installed in his Morane Parasol in early 1915, Garros proceeded to shoot down three German aircraft with comparative ease who went into an attack unaware of the front gun installed in his plane. However, in April 1915 he was forced down behind the German lines and captured before he had time to set fire to his plane. Now the Germans had the gun. Quick to realize its possibilities, it was turned over to Tommy Fokker to experiment with. He was given two days to produce a revised version that could be installed in the *Eindekker*. Calling in two of his most able technicians, they worked round the clock and well within the time limit a vastly improved model was ready. It passed all the tests with flying colors and the order to go full speed ahead was given to equip the *Eindekker* No. 2 with the gun. Both the British and the French had turned down Gilbert's device as lacking reliability and gave it no further consideration.

For months the *Eindekker* was a scourge in the skies until finally months later, a German pilot got lost and came down behind our lines. Both pilot and plane were captured and the secret of the gun revealed. As rapidly as possible the Allies equipped their planes with a still better version. The first British plane to be fitted with the new gun – a Vickers – was the Sopwith 1½ Strutter which came into service in May 1916. It was extremely successful (See below).

Incidentally, Fokker first offered services to the British prior to the war but they refused, which proved to be a sad mistake. At that time, however, he had not shown himself to be in any way superior to our own engineers. Two years later he was offered a substantial sum to swing over to our side, but he was quite satisfied with the situation in Germany and with the many friends he had gathered around him.

Endnotes

1 Anthony Herman Gerard Fokker (1890-1939) was a Dutch airman and pioneer aircraft manufacturer.

2 Roland Garros (1888-1918) became one of the earliest fighter pilots in the war. He developed the first single-seater fighter plane equipped with an on-board machine gun that fired through the propeller. Lacking a synchronizer, he had to armor the propeller with steel wedges to deflect bullets that would otherwise hit the wooden propeller.

Below: Sopwith 1½ Strutter. Courtesy Aeronaut.

Chapter 15: Shot Down in Flames

Through the dark days from the Fall of 1916 until May 1917, the loss of air supremacy by the R.F.C. on the Western Front was a calamity. Not only in the loss of life and planes destroyed, but the manner in which those who flew met their deaths. During this period the headlines in the newspapers printed in the British Isles would blazon forth with the words "Shot Down in Flames." The press seemed to revel in those four words. The appeal to the public was strong as they visualized these intrepid birdmen heroes of the air fighting to the last as the flames raged fiercely around and through them – like the lad who "stood upon the burning deck."

Fighting the flames to the last, a stock phrase so commonly used, no doubt softened the blow to those nearest and dearest to them. But the reality was that of a man pinned down in his seat writhing and screaming in his last agonies as he struggled to get free. Vividly described by Richthofen, his greatest thrill as he repeatedly stated was to follow the flaming plane as it tumbled its way to earth and to watch and gloat over the tortured pilot as the fire consumed, with a fierce heat, another one of his many victims.

Escape was hopeless. The gas tank was immediately in front of the pilot, separated from him by the instrument board and a sheet of metal optimistically described as a firewall. This, theoretically, held back the flames when the gas tank took fire. Overhead and in between the two upper wings and above the engine was the center section reserve tank which filled the main tank using a wobble pump operated by the pilot. This was bad enough, but one plane design went still further by placing a gas tank underneath the pilot's seat. Altogether, this arrangement of tanks materially assisted in sparking the flames that destroyed the planes and those who flew them. But actually there was no other suitable spot where they could be fitted, a condition that also applied to enemy aircraft. When the Americans arrived on the scene they quickly dubbed the R.E. 8 and the B.E. 2c "Flaming Coffins" but flew them all the same and other types were almost as bad.

The greatest fear any pilot had was death by fire and many who suffered from a too vivid imagination in contemplating a situation such as this lost their nerve and tossed in the sponge. Pilots, for all their happy-go-lucky nature, had a certain grim humor which had a marked effect on the young pilots suspected of being far from enthusiastic about the dangers inherent to flying across the line. Whenever Richthofen was mentioned, the conversation would switch to the many and varied forms of death that were inevitable and for which there was no escape, with particular reference to that of fire. One such chap, after listening with horror and without ever having crossed the line, sent in his resignation and returned to his regiment, preferring a bullet in the head to burning up alive.

The risk of fire when starting up the R.E. 8 had been conquered finally but in combat they were just as vulnerable. Richthofen recognized this and trained his pilots to maneuver their planes so that their fire would be directed to the gas tank and engine, and lastly the pilot. There was no escape from the flames and he preferred they die that way rather than from the mercy of a bullet.

The fuel was extremely volatile, and as the gas pouring out from the punctured tank was driven back by the force of the air, it soaked through the clothing and into the pores of the unfortunate victim. Gulping for air, he swallowed the fumes and then as he caught fire his screams were quickly silenced as throat and lungs were ignited and death came swiftly to end his agony. But back behind the pilot was the observer, immobile and helpless, and without aid of a parachute, he waited to be swallowed up in the raging furnace or jump to a more merciful death.

Consumed by Fire

Portrayed clearly are the scattered remains of a plane and the shape of a man shot down in flames – all that was left of the famous German ace, Immelmann. Intact only are the steel cylinders of the Oberusal rotary motor.

On rare occasions a pilot was able to escape the fury of the flames at the first signs of fire by side-slipping the plane with nose held high verging on the stalling point. The side pressure thus created forced the flames upward and outward and clear of the pilot. This also reduced the downward speed of the plane to a mere 50 mph, diminishing the forced draft fanning the flames proportionately.

The danger was the destruction by fire of the upper wings, but given low enough altitude, skillful piloting and an iron nerve, this could be overcome and the plane brought down with none too serious injury to pilot and observer. The plane, of course, would burn itself out but that was of little consequence. Even the planes fought hard for their "lives" and with as great an ingenuity as their pilots.

There were times when the fuel worked with weird effects and more horrible because it was so totally unexpected. A tank can be punctured repeatedly and still not catch fire if the bullets pass through the liquid. This we demonstrated on the ground by riddling a full tank of gas with machine gun fire after learning about the following disastrous ending to what appeared to be a narrow escape from being shot down in flames.

An R.E. 8 was attacked by a flight of enemy albatross and although their gas tank was badly punctured and both pilot and observer were soaked to the skin with gasoline – it was summer and they were flying without the protection of leather jacket or thigh boots – the pilot made his way down safely to the ground and taxied up to the Tarmac in front of the hangar. Half-blinded, they climbed out of the plane and almost as soon as they were clear both men burst into flames. As they rolled in agony on the ground the mechanics, with the aid of fire extinguishers, were able to put them out. But it was impossible to touch them as they were burning inwardly from the fumes they'd swallowed. They lay there through the night and it was not until the early hours of the following morning that it was possible to move them, by which time the smoke no longer curled up in thin wisps from their charred bodies which crumbled as they were lifted.

Shot down in flames, many did not burn until long after the scrap was over when the slightest friction would set off the fatal spark and then no power on earth could put out the flames. On the other hand, many were drenched with gas and nothing happened. In such wrecks where the plane was totally destroyed, the pilot had hardly been touched by fire whereas the lower part of the other fellow's body was burnt beyond recognition .

Right: The wreckage of Immelmann's Fokker E.III 246/16 at its crash site. The tail section and wings broke away in midair and fluttered to the ground separately. Immelmann was killed in June 1916. Courtesy of Aeronaut.

Above: This aircraft "made the rounds" in order to familiarize nearby *Jastas* with the new biplane fighter. Here it is pictured on the airfield of *Jasta* 14 at Phalempin on 23. May 1918, the Triplanes operated by *Jasta* 14 can be seen in the distance. Also stationed at this airfield at this time, but not in the photo, was *Jagdstaffel* 30, which was then operating the Pfalz D.IIIa.

Chapter 16: The Lord Chamberlain Commands

On July 29, 1917, a field message came through to the C.O. No. 5 Squadron as follows:

To – O.C. No. 5 Sqdn – Sender's Number M 591
Under authority by H. M. the King C. in C. Awarded Military Cross to Lieut. R. Duncan, 1st Can. Mtd. Rifles and No. 5 Sqdn. R.F.C.

G.O.C. 1st Bde. & Wing Commander send Lt. Duncan their congratulations.

From 1st Wing

This was the first notification of my being awarded the Military Cross. This is a decoration only awarded men of commissioned rank and except under unusual circumstances, it has to be received from the hands of the King himself which, in my case, was several months later.

To say the least, I was pleasantly surprised as I'd never given a decoration a second thought. But I was striving hard for promotion to Flight-Commander which meant a step up in rank and more pay. Anyhow, the winning of it was the accumulation of several good jobs I'd done as General Shepherd informed me when he flew over to offer me his personal congratulations. A few months later he was killed on the Western Front much to the regret of us all in the R.F.C.

In December 1917, I was posted to the Home Established, which means transferred for a tour of home duty. Three days after I arrived in London on December 5, 1917, I received the following wire:

O.H.M.S. Buckingham Palace
To: Capt. Richard Duncan, Bank of Montreal Waterloo Place
Your attendance is required at Buckingham Palace on Wednesday next the twelfth inst at ten o'clock a.m. Service dress please. Telegraph acknowledgement.

Lord Chamberlain, London

The Bank of Montreal was the official address for most Canadian officers. On leave we would furnish them with a forwarding address. Mine, at this time, was the Cavendish Hotel, Jermryn Street, London, owned by the famous Mrs. Lewis.[1]

I got the wire in the morning and after reading it through decided I didn't want to waste valuable leave time messing around at the Palace. So I drove over to the war office in Whitehall to find out if, rather than go to the Palace, I could have the medal sent to me c/o The Bank of Montreal? The answer was an emphatic "No." Also, that above all else, I be there on time. I really believe the staff officer I talked to thought I was off my rocker, but I really had a dislike for pomp and circumstance.

The night before I was to report at the Palace I left word with the porter at the hotel to call me at 8:00 a.m. At 10:00 a.m. I woke up, nobody had called and here I was in my hotel room instead of standing by waiting for the King. Working against time as I did and rushing around like mad, I did not arrive at the Palace until 10:45 a.m.

At the beginning of a long flight of broad steps a footman resplendent in knee breeches and heavily brocaded coat took my hat and cane and ushered me up the steps into a huge chamber beautifully furnished in which colors of red and gold predominated. In my left hand I carried my gloves according to regulations

There were about thirty other officers standing in a loose line according to rank, each with hook pinned to his chest and among them was my cousin, Captain Torn Webster, 29th Winnipeg Battalion. We had been out the night before and he told me he was determined to witness the ceremony come what may. So he had "trashed" his way in, one of the few who ever got away with it, expecting to see me there, and as I failed to show up when the officers began distributing the hooks, he grabbed one ready to stand in for me if I failed to appear. Nobody could have been more surprised than myself to see him standing there ready to keep on "crashing" and meet the King.

A few minutes later the King appeared on the scene. Torn gave me his hook and I, when my turn came, walked into a smaller room where the King stood flanked by high ranking officers of both services. One held a velvet cushion on which the decorations rested and as I advanced to meet the King, read off the citation. Then the King shook me by the hand, said a few words of congratulation and hung the medal on my chest. I thanked him, walked back two paces, turned and passed through a door to my right. There a footman took the military cross, fitted it into a royal purple leather covered case with a golden crown on the top and that was the end of the ceremony. He handed it to me with a pleasant smile and a murmured "good luck."

> But my pride was tempered as swiftly through my mind passed the picture of the men who had fought so gamely to help me and lost so much themselves – three of them would never fly again, one of them was the Australian returning to his distant home, shattered in nerve and little aware of the world around him. They were the bravest of the brave, my observers, and they are the men who rightly earned this medal I received from the King.

Endnotes

1 Mrs. Rosa Lewis, who bought The Cavendish in 1902, put her husband Excelsior Tyrel Lewis and his sister Laura in charge. When they failed to maintain it properly, she took charge, expanded it to over 100 rooms, and made it into a much favored dining spot for King Edward VII. During the war, she turned the hotel into a "social first aid center" for servicemen. She did the same in World War II. In 1962, the Cavendish closed its doors and was demolished.

Above: Sporting the full factory paintjob and standard armament, Fokker E.V 133/18 is seen here after delivery to the front. The plane was accepted at Schwerin on 30 July, its factory works number was 2774. This number is clearly visible on the forward fuselage, just behind the middle "tripod" strut. (Tobias Weber)

Above: Lieutenant R. Duncan. Courtesy of the R. Duncan Collection at Southern Illinois University at Carbondale.

Right: Military Cross. Courtesy of Wikimedia Commons.

Endnotes

1 The term "N.F." referred to hunting for artillery guns that were firing.

2 The Queen's Westminsters was an infantry regiment of the Territorial Army. Designated a Reserve Force, it originated in New Westminster, British Columbia in 1910.

3 The "Mademoiselle from Armentiéres" is an English song written by Harry Carlton and Joe Tunbridge and a variation of "Hinky Dinky Parlez-vous" and tells of an encounter between an inn-keeper's daughter and two German officers.

4 The Battle of Arras was a British offensive that lasted from April 9 to May 16, 1917. Searching for a breakthrough on the Western Front, it resulted in the furthest advance since the beginning of trench warfare. The British advanced on a broad front between Vimy in the northwest and Bullecourt to the southeast. The Canadian Corps of the 1st Army captured Vimy Ridge. Eventually, it ended in a stalemate for both sides.

5 In 1917 the German Air Army Service created the Jagdgeschwader fighter group and nicknamed the "flying circus" because of the brightly colored aircraft that they flew.

6 The Military Cross (MC), is a third-level military decoration established by King George V on December 28, 1914, and awarded to captains or officers of lower rank up to warrant officers for acts of exemplary gallantry during active operations.

Left: French Voison 8 night bomber. France and Britain relied mostly on single-engine pushers for night bombing; in contrast the Germans used twin-engine night bombers. Courtesy Aeronaut,

Below: French Voison 10 night bomber; note dark finish and the landing lights on the front of the nacelle. Courtesy Aeronaut.

Made in United States
North Haven, CT
06 November 2026

21791729R00050